G. A. JONES

The Properties of Nuclei

SECOND EDITION

CLARENDON PRESS · OXFORD

1987

Oxford University Press, Walton Street, Oxford OX2 6DP

Oxford New York Toronto
Delhi Bombay Calcutta Madras Karachi
Petaling Jaya Singapore Hong Kong Tokyo
Nairobi Dar es Salaam Cape Town
Melbourne Auckland

and associated companies in
Beirut Berlin Ibadan Nicosia

Oxford is a trade mark of Oxford University Press

Published in the United States
by Oxford University Press, New York

British Library Cataloguing in Publication Data
Jones, G. A.
The properties of nuclei.—2nd ed.—
(Oxford physics series)
1. Nuclear physics
I. Title
539.7′4   QC776
ISBN 0-19-851868-4
ISBN 0-19-851869-2 (Pbk)

Library of Congress Cataloging in Publication Data
Jones, G. A. (George Arnold)
The properties of nuclei.
(Oxford physics series; 12)
Bibliography: p.
Includes index.
1. Nuclear structure.   2. Nuclear reactions.
I. Title.   II. Series.
QC793.3.S8J66   1987   539.7′23   86-17931
ISBN 0-19-851868-4
ISBN 0-19-851869-2 (pbk.)

Typeset and Printed in Northern Ireland by
The Universities Press, Belfast

Editor's Foreword

This book fits closely into the network of topics of the Oxford Physics Series and at the same time preserves an integrity of its own. The earlier complementary core texts include *Radiation and Quantum Physics,* and *Atomic Nuclei and their Particles* with supporting texts on *Electromagnetism* and *Special Relativity.*

The volume *Basic Quantum Mechanics* will provide the necessary introduction to many sections of this new text on *The Properties of Nuclei,* the level of which is pitched to take over where the corresponding sections of *Atomic Nuclei and their Particles* leaves off.

Dr Jones concentrates his attention on the available data and their interpretation, developing a natural sequence from the properties of nuclear forces, through nuclear models to nuclear decay and reactions including fission and fusion. The study of these topics is undertaken in second- or third-year courses in physics, applied physics, and nuclear engineering and challenges the student to appreciate both the power of the available experimental and theoretical methods and also the wide range of opportunities for research in the nuclear many-body problem. The sureness of approach found in this treatment illuminates the scene in a rewarding fashion which reflects the author's experience of teaching and developing the subject over a number of years.

March 1977   E. J. B.

Preface to the Second Edition

In the second edition the contents of the chapters has in the main remained unaltered except in Chapter 6 where the topic of Coulomb Excitation has been introduced as an example of direct interactions. The chief changes in these chapters consist of amplification of the text in places where teaching experience has indicated that this was desirable.

Extension of the subject matter is contained in three of the four new Appendices. The transition between isolated resonances and continuous cross-sections has been touched upon in the original text, but an appendix now treats it in more detail. Similarly Partial Wave analysis, also referred to obliquely in the text, is now discussed in somewhat more detail. Although this approach to cross-sections is more 'natural' than summing over resonances, the mathematical techniques required for its full appreciation are, unfortunately, beyond the scope of the average reader of a book at his level, so this appendix merely indicates the way this topic develops. An important addition, in two of these appendices, concerns the influence of isospin on nuclear transitions which illustrates its importance in nuclei and should lead students toward a better understanding of its influence in the interactions of elementary particles.

I am very much indebted to Dr P. E. Hodgson who read a first draft of this edition and made many useful comments which were incorporated into the final draft.

Oxford
May 1986

Preface to the First Edition

The level I have aimed at in this book is that of the nuclear-structure course given to Oxford students in the final term of their second year. In their third year some, but not all, pursue a more advanced course, according to their choice of options. The book therefore reflects the requirements of 'mainstream' undergraduate physics courses. It presupposes a certain familiarity with particle accelerators and radiation detection.

The content is chiefly devoted to the static and quasi-static properties of nuclei and nuclear excited states. But I firmly believe that all physics students should have some notion of how a nuclear reactor works, and so a chapter on nuclear reactions has been included in addition to a brief account of fission (and fusion).

In a book of this length much has to be taken on trust; quantum-mechanical formulae have been used freely with no more than a passing reference (the particular reference is to a text which I found useful as a student and still do as a teacher, but a more recent publication may appeal to the reader). Whatever the choice, I hope my book will encourage the student to acquire a working knowledge of basic quantum mechanics, which forms the framework of his (her) studies.

I am indebted to members of the Department of Nuclear Physics, Oxford, who read and commented on the script, and especially to Dr M. G. Bowler with whom I had many valuable discussions. I should also like to thank one of the editors of this series (E. J. B.), who, I feel, put more time than his editorial duties demanded into a critical appraisal which I much appreciated.

G. A. J.

Contents

1. General properties of nuclei

Introduction

From early measurements on Coulomb scattering it is known that the atomic nucleus is the small particle at the centre of the atom (radius $\ll 10^{-11}$ cm) which possesses nearly all the mass of the atom and which carries a positive charge equal to the atomic number of the atom. From later work it is known that the nucleus is built up of protons and neutrons; the former supply the positive charge, whilst both contribute to the mass. From general quantum-mechanical considerations such a confined system will possess energy levels characterized by quantum numbers. Under normal circumstances, the nucleus will be in its ground state, and this chapter is concerned with nuclear properties in this state. To determine the properties of the nucleus in other (excited) states it is necessary to set up such states by means of nuclear reactions.

Thus a particular nucleus (in its ground state) must be described in terms of the quantum numbers 'angular momentum' (also loosely called spin) and 'parity' (to be explained later in the chapter). This is the minimum number of quantum numbers required by the symmetry properties of the system. There are others, but they are approximate or model-dependent, i.e. they depend on a simplified picture of the nucleus, and, unlike the corresponding atomic system, no one model stands out sufficiently to attach extra significance to its quantum numbers.

In addition, a system of finite extent can be described in terms of moments of the matter and charge distributions. In the case of the nucleus these moments are usually determined from its interaction with the atomic electrons which is electromagnetic in origin, so the moments measured are electric and magnetic multipole moments. A system can have an electric dipole moment only if it has an up–down directional asymmetry in space, e.g. the molecule HCl, but not Cl_2, can have a permanent electric dipole moment. For nuclei no such asymmetry is believed possible, so nuclei do not possess such a moment. Since the components of the nucleus, protons and neutrons, have intrinsic spins and magnetic dipole moments, the nucleus itself can also have a magnetic moment. The two important moments of the nucleus, apart from the electric monopole, are therefore its magnetic dipole and electric quadrupole moments.

Whilst nuclear physics deals with all nuclei, of especial interest are the so-called 'stable nuclei'. These are the nuclei which exist in the world around us, and this is quite specific, though 'stable' itself is much less definite. A nucleus may be unstable against emission of a neutron or a proton, of a composite particle such as an α-particle, or against β-decay. Of these neutron emission is very fast ($\tau \sim 10^{-16}$ s or less), proton emission generally less so but still very fast, whilst the latter two can have lifetimes varying over a vast range depending on the energy available ($\tau \sim \mu$s to $\sim 10^{20}$ years). By 'stable' a lifetime of the order of the age of the universe ($\sim 10^{10}$ years) is implied; sometimes it has the restricted meaning 'stable against the β-decay process'.

The size

The original measurements made by Rutherford and his co-workers have now been augmented by more accurate work, which has not only measured an RMS radius for the nucleus but has also obtained a measure of the rate of fall-off of matter density, and charge density, at the periphery. The measurements fall into two classes: (1) those that determine the charge distribution and (2) those that determine the matter distribution.

The charge distribution

These measurements are made with charged leptons as probes. Leptons form a class of fundamental particles which interact only weakly with protons and neutrons, except for the long-range Coulomb term where relevant.† Thus the interaction between an

---

† Interactions between fundamental particles can be put into the categories strong, electromagnetic, weak, and gravitational, which are roughly in strength as $10:1/137:10^{-7}:10^{-39}$ (see e.g. Perkins 1972).

electron or a muon and nuclear matter is completely dominated by the Coulomb interaction. The greatest source of information is the elastic scattering of electrons. Accurate measurements of optical and X-ray spectra and muonic X-ray spectra have also given information on the nuclear size.

Electron elastic scattering

Outside the nucleus of charge Ze the electron experiences a force $\propto Z/r^2$, but inside the nucleus (assumed spherical) the force is $\propto Z'(r)/r^2$, where $Z'(r)$ is the charge contained in the sphere of radius r. Thus small-angle scattering, which corresponds to a classical trajectory rather distant from the nucleus, will obey the Rutherford formula and therefore give no information about the distribution of charge, whilst large-angle scattering will show a significant deviation from the formula. The ratio to Rutherford scattering as a function of angle is determined by the form of $Z'(r)$ and therefore by the charge distribution. From the uncertainty principle, if we wish to attach significance to changes of density over a distance Δr, then the incident particle must be capable of transferring momentum $\hbar q$ to the target nucleus such that $q\Delta r \sim 1$; obviously the wavenumber of the incident beam must exceed q. For $\Delta r \sim 10^{-13}$ cm, electrons of kinetic energy >200 MeV are required. Some twenty years ago, measurements

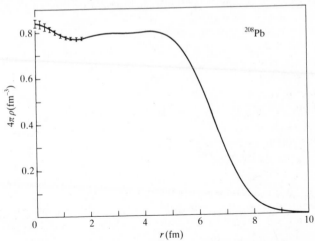

Fig. 1.1. The radial charge density of $^{208}_{82}$Pb from the analysis of a number of electron-scattering experiments and incorporating the results of muonic X-ray measurements, (Based on Friar and Negele (1973). *Nucl. Phys.* **A212**, 93.)

were made at 50 MeV and were capable of defining a suitably averaged charge radius only; more recently, the energy range has been extended to 600–1000 MeV and the radial dependence of the nuclear charge distribution has been revealed in fine detail, see Fig. 1.1.

X-ray and optical spectra

Since the nucleus has finite extent its field of force on the atomic electrons will not be proportional to $1/r^2$ when the electron is within the nucleus. If therefore the atomic energy levels can be calculated accurately for a point nucleus and compared with their measured values, the nuclear size can be deduced. But the calculation is an unsolved many-body problem. For X-ray transitions the screening effect of the outer electrons is not large and can be approximated to; but for the tightly bound inner electrons there are uncertainties even in the solution of the two-body problem, whilst for the valence electrons, although these latter uncertainties do not arise, the screening effects cannot be calculated to sufficient accuracy. In certain cases useful information can be derived. It is reasonable to assume that, for a comparison between spectral lines from isotopes of the same element, the differences will arise from the differences in nuclear mass and size only. For light elements the mass difference produces effects large enough to measure, but for heavier elements the measured effects must be ascribed to differences in nuclear charge distribution. Thus the method of isotope shift gives information on the way the RMS radius of the charge distribution changes on adding neutrons to a nucleus.

Muonic X-rays

Negative muons μ^-, like electrons, can be captured by atoms, and interact electromagnetically with nuclei. Since they have a mass $\sim 200\, m_e$ (where m_e is the mass of the electron), their Bohr orbits will have 1/200th of the radius of the equivalent electron orbits. For a heavy element such as lead, the lowest Bohr orbit (since no orbits are occupied by other muons, the exclusion principle does not operate here) is such that the muon spends most of its time within the nucleus. The size effect is therefore large, reducing the lowest $p \rightarrow s$ transition from ~ 16 MeV for a point nucleus to ~ 6 MeV as measured. The muon is, of course, unstable and decays spontaneously to an electron and two neutrinos (v) with a mean life of $\sim 2\,\mu s$. In the lowest Bohr orbit in lead it decays more quickly because of the competing mode $\mu^- + p \rightarrow n + v_\mu$, but during its lifetime it transverses several metres of solid nuclear matter (density $\sim 3 \times 10^{17}\,\mathrm{kg\,m^{-3}}$), thus emphasizing the weakness of the weak interaction.

From a theoretical aspect, the computational difficulties are the same as those encountered in interpreting electronic X-rays, but the uncertainties are reduced. Obviously the screening effects of the atomic electrons will only be of the same order as for the K-electrons, whereas the nuclear interaction energy is increased by the factor 200. It also turns out that the uncertainties in the two-body problem are reduced by the same factor. Since the bound muons have effectively a long wavelength ($\sim$ size of the Bohr orbit) the measurements lead to an average nuclear radius and give no information on the charge distribution. But the average they give is very precise, and nowadays it is used as a datum in the analysis of electron scattering, acting as a constraint when looking for the finer details of the charge distribution.

The matter distribution

If the nuclear probe interacts strongly with protons and neutrons then the size determined from the interaction will be that of the nuclear matter as a whole. We have also the added complication that the strong interaction has a range which, though short, is not short compared to the nuclear size, and due allowance must be made for this interaction when deducing the mass distribution.

The historical example of this kind of measurement is the elastic scattering of α-particles, which led to the first determinations of the nuclear size. Other charged particles have since been used, e.g. protons, carbon and oxygen nuclei (in the form of partially stripped ions), and other heavy ions. From a theoretical point of view, the simplest process is the scattering of neutrons, but experimentally precise measurement is difficult, since it involves three nuclear interactions—one to produce the neutron, another to scatter it, and a third to detect it. Other methods involving the electrostatic term are of historical importance: they are the lifetime for α-radioactivity and the determination of the electrostatic energy of the nucleus from the energy difference between mirror pairs.

Scattering of charged particles

Experimentally, the scattering cross-section (see Appendix A) is measured as a function of angle for a fixed beam energy or as a function of energy at a fixed angle. In either case the ratio of measured cross-section to Rutherford cross-section is plotted as a function of the variable (see Fig. 1.2). The ratio will be unity over a range of angles from near zero up to a certain angle, beyond which it drops off, indicating absorption of the incident particle. Alternatively, for constant angle the ratio will be unity up to a certain energy, beyond which it drops off. However, for protons or even α-particles on light nuclei it is possible for the cross-section to rise above the Rutherford value, because such particles stand a good chance of being emitted after absorption. For heavier ions, or protons and α-particles on heavy nuclei, emergence of the absorbed particle is improbable. Even so the wave nature of the process modifies the classically expected result (see the next section below) as shown in Fig. 1.2.

The region of fall-off is correlated with the classical trajectory passing through the periphery of the nucleus, immediately giving an idea of the size of the nucleus. How to obtain a more precise determination of the size was uncertain before the development of the optical model (see Chapter 3). We should notice that, in

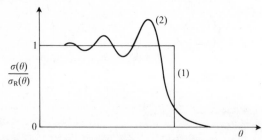

Fig. 1.2. Ratio of elastic scattering to Rutherford scattering for a charged particle (heavy ion) on a medium size nucleus, at an energy (in the centre of mass system) which is higher than the Coulomb barrier. (1) represents classical scattering from a spherical nucleus with a sharply defined radius. The angle at which $\sigma(\theta)$ falls to zero corresponds to a classical trajectory which just grazes the nucleus. (2) represents quantum mechanical scattering from the same nucleus. The modulation is of the form of Fresnel, rather than Fraunhofer, diffraction because the Coulomb deflected particles appear to emanate from a region close behind the nucleus. For a more realistic nucleus with the density falling off to zero over a range of the nuclear radius, and a similar absorption function, the diffraction pattern is reduced by 'blurring'.

the interpretation of the results, in addition to making allowance for the range of nuclear forces it is also necessary to take into account the finite size of the incident particle. Initially it was felt that heavy ions with their strong absorption should best yield to theoretical treatment, but more recently, with the optical model and its parameters well established, the analysis of elastic scattering of 800 MeV protons has provided the most accurate information available on the matter distribution.

Neutron scattering

When fairly energetic neutrons ($\sim$10–30 MeV) impinge on nuclei they are strongly absorbed from the initial beam, though they may subsequently be emitted from the resulting excited nuclei. The degree to which they are emitted depends upon the number of competing modes of decay (see Chapter 6). If there are many such modes, and there are for $E \sim 10$–30 MeV, elastic scattering will be reduced to a minimum. Elastic scattering cannot be zero since an extension of Babinet's theorem in optics to this nuclear wave problem indicates that absorption without scattering is impossible. This scattering is predominantly through small angles (up to $1/kR$ as in optical diffraction). The absorption cross-section can be conveniently measured by a beam-attenuation technique. A neutron detector placed in the path of a collimated neutron beam determines the beam flux before and after interposition of the absorbing target. If the angle subtended by the detector at the target is greater than the angle of diffraction, then diffracted neutrons are detected, and the beam attenuation is a true measure of nuclear absorption. A simple theory gives $\sigma_{abs} = \pi(R + \lambda)^2$, see Appendix H, where the additional term λ, the wavelength of the projectile, expresses the uncertainty with which we can locate a neutron. A more sophisticated analysis can be carried out in terms of the optical model. Although in principle this technique should be equivalent in usefulness to proton scattering, at least when taken to higher neutron energies, in practice the added difficulties in producing neutron beams and detecting scattered neutrons make for much reduced precision.

α-Radioactivity

From consideration of the mass equation (see pp. 11–17), all nuclei above $A \sim 150$ are unstable, in principle, and can break up into two smaller units. Because of its high intrinsic binding, the α-particle is usually one of these units. The process involved will be discussed in more detail in Chapter 4, but for the present it can be summarized by stating that the probability of α-particle emission depends on two factors: the intrinsic nuclear factor and the barrier penetrability. The latter is a rapidly varying function of the nuclear radius, which can therefore be determined if the former is known—the measured lifetime is known with great precision. Unfortunately, the intrinsic factor is a notional one and cannot be measured; theoretical estimates are uncertain, with the result that the nuclear radius deduced is qualitative only. It is probably more meaningful to use the data, and a nuclear radius from other data, to deduce the intrinsic nuclear factor.

Electrostatic energy differences of mirror nuclei

Mirror nuclei are nuclei in which the proton number and neutron number are interchanged, e.g. ^{13}C and ^{13}N having $Z = 6$, $N = 7$ and $Z = 7$, $N = 6$. Most of the measurements refer to mirror pairs having $|N - Z| = 1$.

It is believed (Chapter 2, p. 33) that the specifically nuclear interactions will be the same for such pairs, so their mass difference will arise from the electrostatic interaction and the n–p mass difference. The former term is discussed in greater detail in the next section; for the moment a uniform distribution of charge in the nucleus is assumed to give for the mass difference

$$\Delta M = M(A, Z + 1) - M(A, Z)$$

$$= m_p + m_e - m_n + \frac{3}{5}\frac{1}{4\pi\varepsilon_0}\frac{(2Z + 1)e^2}{R},$$

where the M and m are defined in the next section.

The mass difference can be measured accurately from the β^+-decay spectrum of $(A, Z + 1)$, since

$$\Delta M = 2m_e + E_{kin} \quad \text{—see eqn. (1.3a)},$$

where E_{kin} is the maximum kinetic energy available to the β^+; or from the threshold energy for neutron production by protons on (A, Z),

$$M(A, Z) + m_p + m_e = M(A, Z + 1) + m_n + Q$$

$$\Delta M = m_p + m_e - m_n - Q$$

Thus the Q-value of the (p, n) process is the negative of the electrostatic energy difference.

From the data, $R = R_0 A^{\frac{1}{3}}$ is found to hold good, but the value deduced for R_0, namely 1.46 fm is rather large compared with deductions from other processes (see the end of this section). This reflects the dubious supposition that the nuclear charge is uniformly distributed within the nucleus. Since the data on mirror nuclei refer only to light nuclei $(A < 40)$ surface effects will be expected to be large.

K-mesic X-rays†

From the discussion of muonic X-rays, it follows that the K^--mesic Bohr orbits will have an even smaller radius. In addition, the K^--meson will interact strongly with the nucleus, so nuclear absorption will compete with X-ray emission. Absorption will occur from highly excited Bohr orbits when the K-meson overlaps the tail of the matter distribution. These measurements therefore focus attention on the details of the matter distribution at the surface of the nucleus.

Initially it was hoped that the measurements would reveal the degree of clustering in the 'nuclear stratosphere' since an interaction between K^- and a single proton or neutron results in the production of a pion, whereas the interaction of a K^- and a close pair of nucleons (as in perhaps an α-cluster inside the nucleus) does not produce a pion. In the event, the analysis has been hindered by the lack of knowledge of the K^- interaction at low energies—it has turned out to be more complex than it was initially thought to be. Concerning the interactions with single nucleons, the charge differences enable a distinction to be inferred between neutron and proton and indicate a preponderance of neutrons in the stratosphere; but this is to be expected since protons are additionally inhibited by the Coulomb barrier in this region where both protons and neutrons have negative energies, i.e. they are virtual particles.

Conclusions on size

Analyses of the earlier data, which were capable only of defining an effective nuclear radius, gave the result that the radius is

† K-mesons, which can be positively or negatively charged or neutral, have a mass $\sim 1000\, m_e$ and interact strongly with nucleons to produce 'strange' particles.

proportional to the cube root of the mass number A, i.e. that the nuclear density is a constant independent of the nuclear size. Later measurements indicate that, although this is true for the central region, near the periphery the density falls off smoothly and quickly to zero. The precise nature of the fall-off is not determined from the data, but it has proved convenient to represent it by the (Fermi) function

$$\rho(r) \propto \left[1 + \exp\left\{ \frac{(r-R)}{a} \right\} \right]^{-1},$$

where R, the radius at half-density, is referred to as the nuclear radius and a is a surface thickness parameter which defines the region over which the density is falling off. This latter parameter appears to be independent of nuclear size, whilst $R \propto A^{\frac{1}{3}}$ though most recent measurements indicate small deviations from this smooth curve, which can be ascribed to shell effects (see Chapter 3).

The measurements referring to the charge distribution are of greater precision than those referring to the matter distribution and the values for R and a are much the same. In more detail, there seems to be a small but significant difference between R_p and R_n—the latter having been unfolded from the charge and matter distributions. In the larger nuclei ($A > 100$ or so) the radius for the neutron distribution exceeds that for the proton distribution by 0.1 to 0.2 fm, i.e. of the order of 2–4 per cent with a probable error of about 1–2 per cent. In the lighter nuclei the difference falls below the level of significance, except in special cases like $^{18}_{8}O$ with two neutrons outside the 'doubly magic' core of $^{16}_{8}O$, and $^{48}_{20}Ca$ with 8 neutrons outside the doubly magic $^{40}_{20}Ca$ (see Chapter 3 for shell model configurations). The data may be summed up by putting $R_p = R_0 A^{\frac{1}{3}}$, where $R_0 = 1.1$ fm, and $a \sim 0.5$ fm—this latter figure is not known accurately. Neglecting the above small differences the neutron distribution will have the same parameters, but they are less accurately known.

An objection may be raised, by the more discerning student after reading on, that nuclei have been treated as spherical whereas some of the more massive nuclei have quite large quadrupole moments. The reason is that the nuclei have not been aligned in their targets during the experiments described, and so the results give a time- and space-averaged picture of the nucleus which will have spherical symmetry even when it is intrinsically non-spherical. The averaging process may well be expected to produce a wider region of fall-off i.e. a greater value for a. There is little evidence for this either way, partly because a is not known accurately anyway, but chiefly because measurements on nuclear radii tend to concentrate on nuclei near enough to the closed shell configuration to ensure that they are spherical (see Chapter 3). There are good reasons for this both experimentally and theoretically; a non-spherical nucleus has a rotational band of low-lying states based upon the ground state and these states are very easily excited by a charged projectile in a close encounter. The proximity to the ground state makes it difficult to distinguish between elastic and inelastic events experimentally, and the fact that the inelastic process can be a very large fraction of the elastic necessitates the use of different theoretical techniques, known as 'coupled channel' calculations since the elastic and inelastic channels are represented by simultaneous differential equations which are strongly coupled to each other, whereas in the more normal 'weak coupling' calculation the differential equation for the elastic channel can be solved in isolation, the coupling terms being lumped together into a single term representing absorption.

The mass and stability

Historically, the masses of neutral atoms were determined relative to (atomic) $^{16}O \equiv 16$ atomic mass units and later relative to (atomic) $^{12}C \equiv 12\,u$ (unified atomic mass unit), by chemists using chemical reactions. The measurements were later improved upon using physical techniques involving the trajectories of ionized atoms through electric and magnetic fields. With this information available the nuclear physicist found it expedient to use atomic masses in his equations for nuclear reactions, since conservation of charge ensures that the same number of electrons will appear on each side of an equation (though not for β^+-decay, see p. 18). Balancing the electron number, however, does not allow completely for the difference between atomic and nuclear masses, since no account has been taken of the binding energy of the atomic electrons—which, though small, can add up to ~0.5 MeV for uranium. If this appeared directly as an error in the determination of Q-values in nuclear reactions then it would be serious. But most reactions are concerned only with small changes of A and Z and therefore only with small changes in this term, which, if neglected, could introduce errors of order 20 keV for a-decay in the region of uranium when using nuclear masses—though the correction is well known and accurate to a fraction of 1 keV. On the other hand, use of atomic masses largely eliminates such effects since the mass measurements are done on ions which are ionized in the valency shell and thus already include all the large contributions to electronic binding. For this reason the atomic scale is to be preferred. However, the student may come across the usage of nuclear mass and so, although the atomic scale will be employed here, comparisons will be made between the two scales. The nomenclature employed will be that M will refer to atomic masses and m will refer to nuclear masses. Since the masses quoted for the proton and neutron are usually m_p and m_n, rather than M_H ($M_n = m_n$ anyway), the mass equations will contain $Z(m_p + m_e)$ rather than ZM_H.

From the equivalence of mass and energy, the mass of a nucleus consisting of Z protons and N neutrons ($Z + N = A$) will be less than the sum of the masses of the individual nucleons by the energy of binding (expressed in u (1.66×10^{-27} kg)—it is in fact more convenient to express the masses in MeV and since, in u, the mass values are close to the integers A, it is usual to give the mass excess in MeV, i.e. $\{M(A, Z) - A\}$u converted into MeV). Thus

$$M(A, Z) = Z(m_p + m_e) + (A - Z)m_n - B.$$

It is necessary therefore to consider all contributions to the binding energy. This is done in an empirical way, but guided by physical facts and theories, in the mass equation of von Weizsäcker.

From the fact that nuclear density (neglecting surface effects) is a constant a nucleus can be compared with a liquid drop, for which it is known that the binding energy (again neglecting surface effects which are usually small for liquid drops) is proportional to its volume, since all molecules deep in the liquid are in the same environment. Hence, by analogy, the leading term in the mass equation representing the binding energy will be $-\alpha A$, where α is some undefined constant. All constants introduced into the mass equation will be defined to be positive.

A second term in the binding energy also arises from the analogy with a liquid drop. The forces which keep the liquid drop together (van der Waals forces) are short range, so that the volume binding energy term is dominated by the interactions between near neighbours. At the surface a molecule will have fewer near neighbours than a molecule in the interior and so will be less tightly bound, giving a surface-energy term and a surface tension, the former being proportional to the surface area. It results in a lower total binding of the drop and therefore appears as a positive term in the mass of the drop. Since the surface area of a sphere is proportional to r^2, and, reverting to the nucleus, $r \propto A^{\frac{1}{3}}$ the surface-energy term will appear as $+\beta A^{\frac{2}{3}}$ in the nuclear mass equation. As a large fraction ($\sim\frac{1}{2}$) of nucleons are near the surface this will be expected to be a considerable fraction of the volume term.

The anology with a liquid drop neglects the charge on the nucleus. On attempting to remedy this by charging the drop, the analogy breaks down, since the charge migrates to the surface of the drop whereas, from the previous section, it is known that the charge density of a nucleus is proportional to the matter density. This illustrates the difficulties in making classical models of quantum-mechanical systems; obviously the uncertainty principle operates to reduce the energy when both protons and neutrons make full use of the space available. Assuming that the nuclear charge is uniformly distributed over the volume of a sphere then its electrostatic potential energy can easily be shown to be $\frac{3}{5}(Ze)^2/4\pi\varepsilon_0 R$. If the discrete nature of the charge on the proton is taken into account, perhaps Z^2 should be replaced by $Z(Z-1)$. This refinement is probably an improvement but, except for small nuclei, it makes little difference in the mass equation. In any case there are larger factors of ignorance: the charge distribution is known to vary with radius within the nucleus, and there could be a correlated motion which allows protons to avoid each other yet retains the required charge distribution. The electrostatic term is therefore introduced into the mass equation as $+\varepsilon Z^2 A^{-\frac{1}{3}}$, where the (unknown) constant ε is used to account for these uncertainties.

At this stage the mass equation is

$$M(A, Z) = Z(m_p + m_e) + (A - Z)m_n - \alpha A + \beta A^{\frac{2}{3}} + \varepsilon Z^2 A^{-\frac{1}{3}}.$$

Minimizing this function with respect to Z at constant A gives, for stable nuclei, the approximate result

$$Z = A^{\frac{1}{3}}(m_n - m_p - m_e)/2\varepsilon.$$

The important factor is $Z \propto A^{\frac{1}{3}}$, which is far too slow a variation with A (see Fig. 1.3). $A = 32$ has $Z = 16$ stable, so the above formula would give $Z = 32$ stable for $A = 256$, to be compared with the naturally stable $^{238}_{92}U$ having $A = 238$, $Z = 92$.

It is interesting to note from Fig. 1.3 that, up to $A = 40$, the line of stability has $N = Z$. Above this value N increases more rapidly than Z, presumably because of the electrostatic term in the mass equation. It would appear that a term proportional to $(N - Z)^2$, i.e. $(A - 2Z)^2$ is also required, which if of positive sign would mitigate in favour of $N = Z$ for stable nuclei. One need not look far for the origin of such a term; the Pauli principle will

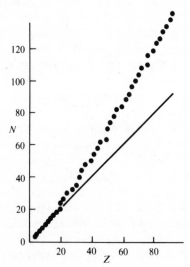

Fig. 1.3. Plot of N against Z of most stable nucleus for given A. For convenience only every sixth A-value is plotted. The full line shows $N = Z$.

ensure that the nucleus (in the absence of the electrostatic term) should have $N \sim Z$ in order to utilize the lowest levels available to both species. To proceed further it is necessary to invoke a model. A convenient model, which incorporates the Pauli principle, is to consider the neutrons and protons as two non-interacting degenerate Fermi gases inside the nucleus. Neglecting the electrostatic term and the n–p mass difference, which are already in the mass equation, the model instantly leads to $N = Z$ for minimum energy. Expanding the energy function from the point $N = Z$ gives a leading term proportional to $(N - Z)^2/A$. Hence a symmetry term $+\gamma(A - 2Z)^2/A$ is added to the mass equation.

An extra term is needed before attempting to use the mass equation to determine which nuclei are stable. In Appendix B it is shown, under approximation, that the balance between terms so far developed in predicting the heavy stable nuclei is so finely poised that the nearest nucleus (of given A) to the line of stability will be most stable. Thus, starting from an even-Z even-N stable nucleus, such that the next stable mass is obtained by adding a neutron, then the second nucleon added to form a stable nucleus will be a proton more often than a neutron since, if (A, Z) is on the line of stability then $(A + 2, Z + 1)$ will be nearer that line than $(A + 2, Z)$. Thus, for any given A, only one stable nucleus will be expected, and of those with even A, stable nuclei will occur as often with odd Z as with even Z.

Nuclear charts show that there are 166 even-Z even-N stable nuclei but only 4 of odd Z, odd N, and these are very light: $^2_1$H, $^6_3$Li, $^{10}_5$B, $^{14}_7$N. Conditions for very light nuclei do not conform to those under which the mass equation was developed, and in addition the systems have few, widely spaced, energy levels which could affect the balance between Z and N. Of the 105 odd-A stable nuclei, 50 are found to have odd Z.

In order to account for this distribution of stable nuclei, a pairing energy is postulated and is accounted for in the mass equation by a small term $\delta(A, Z)$, where δ is a function which has the value zero for all odd A and takes a negative value when both A and Z are even but a positive value when A is even and Z is odd. Its magnitude drops off as A increases, and this is to be expected since, on average, the two similar pairing nucleons will be further apart for large A than for small A.

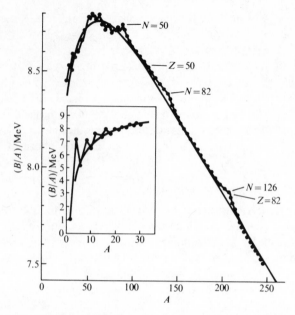

Fig. 1.4. Plot showing the binding energy per nucleon as a function of A. Locations of magic numbers are indicated. The inset continues the plot and illustrates α-particle structure in light nuclei. (Based on Wapstra (1958). Atomic masses of nuclides, *Handb. Phys.* XXXVIII/I.)

The atomic mass equation therefore reads

$$M(A, Z) = Z(m_p + m_e) + (A - Z)m_n - \alpha A + \beta A^{\frac{2}{3}}$$
$$+ \frac{\gamma(A - 2Z)^2}{A} + \frac{\varepsilon Z^2}{A^{\frac{1}{3}}} + \delta(A, Z). \tag{1.1}$$

This five-parameter formula has been fitted to the available mass data (several hundred nuclei neglecting the very light ones, $A < 20$) to give values:†

Parameter	α	β	γ	ε	δ
Value (MeV)	15.6	17.2	23.3	0.7	$11A^{-\frac{1}{2}}$

The fit is shown in Fig. 1.4, where it was more convenient to

---

† The values of the constants vary from source to source. Usually no errors are given so the significance of the variations cannot be assessed. The fits appear to predict the masses of stable nuclei, and their neighbours on the mass diagram, to about 1 per cent in binding energy.

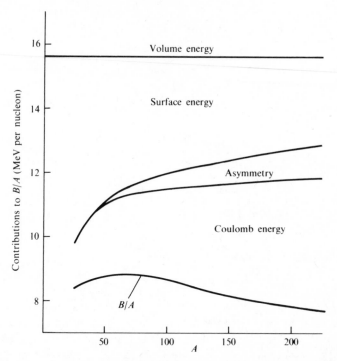

Fig. 1.5. Contributions to B/A of the individual terms in the mass formula.

present the binding energy per nucleon (B/A) against A. To give an idea of the relative importance of the terms, for the stable nucleus $(125, 52)$ the contributions to B/A are $+15.6$ MeV (volume), -3.44 MeV (surface), -0.65 MeV (symmetry), and -3.0 MeV (charge). The variation with A is shown in Fig. 1.5. Notice that, on the scale shown, the pairing energy term is quite negligible; for the neighbouring even nucleus to $A = 125$, it contributes an additional 8 keV per nucleon to B/A.

Stability against β-decay

β^--decay is represented by the reaction

$$(A, Z) \rightarrow (A, Z + 1) + e^- + \bar{\nu}_e,$$

where ν_e and $\bar{\nu}_e$ represent electron neutrino and electron antineutrino (see Chapter 4). This reaction is energetically possible if

$$m(A, Z) > m(A, Z + 1) + m_e \qquad (1.2)$$

since $m_\nu \sim 0$. But

$$m(A, Z) = M(A, Z) - Zm_e$$

leading to

$$M(A, Z) > M(A, Z + 1). \qquad (1.2a)$$

β^+-decay is represented by the reaction

$$(A, Z) \rightarrow (A, Z - 1) + e^+ + \nu_e$$

and is energetically possible if

$$m(A, Z) > m(A, Z - 1) + m_e \qquad (1.3)$$

leading to

$$M(A, Z) > M(A, Z - 1) + 2m_e. \qquad (1.3a)$$

For these two types of decay it seems that the nuclear masses produce a more attractive statement of condition. However, there is another type of decay—an alternative, competing, mode to β^+-emission—namely, electron capture. Nuclei are not required to be stable in isolation, but in their natural habitat when surrounded by atomic electrons. Any one of these electrons can be captured, though if energetically possible K-capture is likely to be dominant, leaving the atom strongly excited. However, in discussing the limit of stability the possibility of capture of the outermost electron, leaving the atom in its lowest state, must be considered, since energetically it is most favoured.

This reaction is represented by

$$(A, Z) + e^- \rightarrow (A, Z - 1) + \nu_e$$

and is energetically possible if

$$m(A, Z) + m_e > m(A, Z - 1), \qquad (1.4)$$

leading to

$$M(A, Z) > M(A, Z - 1). \qquad (1.4a)$$

These equations indicate that electron capture can proceed

when β^+-emission is impossible and also that the atomic mass scale is again favoured: the condition for stability against the β-decay process is that the *atom* (A, Z) should have least mass.

For the heavier nuclei this can be expressed, approximately, as $(\partial M / \partial Z)_A = 0$, which is more convenient to use than the two inequalities. The corresponding condition on the nuclear mass scale $(\partial m / \partial Z)_A = -m_e$, is probably not so convenient to remember nor indeed is the differential condition on the binding energy, $(\partial B / \partial Z)_A = m_p + m_e - m_n$.

From these equations, two nuclei (A, Z) and $(A, Z - 1)$ can both be stable only if they have exactly the same atomic mass. This statement is subject to approximations concerning the neutrino mass and the electron binding energies. Taking these into account might produce a band of stability a few eV wide. The chance that such a pair of nuclei should have equal atomic masses to within a few eV is remote, so it is concluded that, in the few such examples naturally occurring, one of the pair is unstable but has a very long lifetime, e.g. ^{87}Rb and ^{87}Sr—(87, 37) and (87, 38)—of which the former has a lifetime of 5×10^{10} years, emitting β^--particles with maximum energy ~ 274 keV.

Effect of the pairing term on stability

When A is constant and odd the mass values for different Z will define discrete points lying on a parabola. For A constant but even, it is necessary to define two parabolae displaced from each other by $2\delta(A)$, which is roughly 2.5 MeV at $A = 81$, falling to ~ 1.4 MeV at $A = 256$. The odd-Z mass points will fall on the upper parabola and the even-Z mass points on the lower. Figure 1.6 gives examples of the parabolae for odd A and even A.

The displacement of the two parabolae for even A is sufficient to ensure, even when the lowest odd-Z mass point is at the minimum of its parabola, that both its neighbours (which will lie on the lower parabola) have lower mass. This can be checked for the two masses given above by evaluating the second derivative of the mass formula without the δ term; the mass difference between this odd nucleus, assumed to be at the minimum of the parabola, and its two even neighbours will then be $\frac{1}{2}(\partial^2 M / \partial Z^2)_A$. At $A = 81$ this is ~ 1.5 MeV and at $A = 256$, ~ 0.6 MeV, so that the pairing term is amply large enough for its purpose.

These considerations lead to the conclusions that, for odd A,

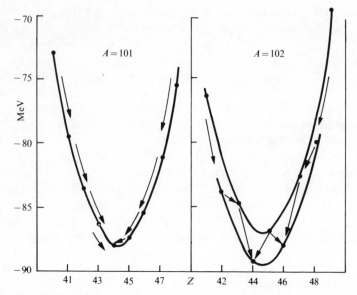

Fig. 1.6. Mass parabolae for odd A and even A. The mass excess (in MeV) is plotted against Z for $A = 101$ and 102.

only one nuclide is expected to be stable against β-decay, but for even A two apparently stable nuclides is likely to be a common occurrence (and is) and even three is a remote possibility (one or two examples are found). The term 'apparently stable' has been used because one such stable even–even nucleus should be able to decay to its less massive even–even neighbour by a double β-decay process, which has in fact been observed, though with great difficulty.

Stability against nucleon or α-particle emission

From Fig. 1.4 it is seen that for $A > 20$ the binding per particle for nuclei near the minima of the mass parabolae (i.e. along the 'valley of stability' on a three-dimensional M, A, Z plot) lies between 6 MeV and 8 MeV. This large value ensures that even the most massive naturally occurring nuclei ($A \sim 240$) are stable against emission of a proton or neutron.

The α-particle, on the other hand, is itself bound by ~ 28 MeV, so that instability to α-emission is much more probable. The condition for instability is

$$M(A, Z) > M(A - 4, Z - 2) + M(4, 2)$$

or

$$B(A, Z) < B(A - 4, Z - 2) + B(4, 2).$$

However, Fig. 1.4 presents B/A as a function of A; if we denote this by $\phi(A)$—*along the line of stability* Z is a function of A, and so $B(A, Z)$ is a function of A only—then

$$A\phi(A) < (A - 4)\phi(A - 4) + 4\phi(4).$$

But

$$\phi(A - 4) \approx \phi(A) - 4 \, d\phi/dA$$

giving, after dividing by 4,

$$\phi(A) < \phi(4) - (A - 4) \, d\phi/dA.$$

From the graph $d\phi/dA \sim -0.75 \, \mathrm{MeV}/100$ and is roughly constant well above the maximum of the curve; and $\phi(4) \sim 7 \, \mathrm{MeV}$.

Thus the condition for α-instability is that $\phi(A)$ is less than

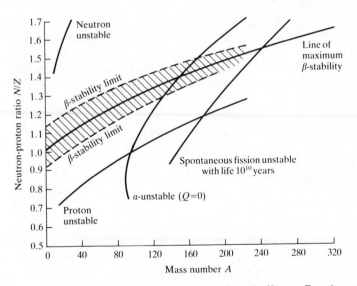

Fig. 1.7. Nuclear stability limits from the mass formula (Segre, *Experimental Nuclear Physics*, Vol. III, Wiley, 1953.)

~7.8 MeV at $A = 100$, ~8.1 MeV at $A = 150$, and ~8.5 MeV at $A = 200$. α-decay is therefore possible for $A > 150$ (though shell effects have an influence; see Chapter 3). Because of the mechanism of the process (see Chapter 4) it becomes a significant mode of decay only above $A \sim 210$. Limits of stability are shown in Fig. 1.7.

Angular momentum (spin) and statistics

Quantum mechanically, as classically, an isolated system has a well defined angular momentum. It is therefore a characteristic property of the nuclear ground state—and indeed any other well-defined state. As such it is loosely referred to as the spin of the nucleus, even though it is a combination of the intrinsic spins and orbital motions of the individual nucleons. The reason for this is probably historical. Nuclear spins of stable and long-lived nuclei were determined from their magnetic interactions with the atomic electrons or with externally applied fields. For such interactions the spin vector is an intrinsic property of the nucleus just as it is, at present, for an elementary particle.

Since the intrinsic spin of a nucleon is $\frac{1}{2}$, whilst all orbital angular momentum vectors are characterized by integral quantum numbers, it follows that the sum of A such spins and orbital terms will lead to an angular momentum (or spin) quantum number J which will be an integer when A is even but (integer $+ \frac{1}{2}$) when A is odd. Examination of a table of nuclear spins confirms this and also reveals a striking fact: all even-Z even-N nuclei have ground-state spin 0, without exception. Since a pairing energy term, favouring the pairing of like nucleons, has already been introduced into the mass equation, it seems reasonable to suggest that such pairs individually couple to zero angular momentum. This point will be taken up in Chapter 3.

The Pauli principle requires that quantum-mechanical systems have wave functions which are antisymmetric to the exchange of like fundamental particles. If the system under consideration is atomic (for example, consider a molecule made up of two like atoms) then the electronic wave function should have a definite symmetry on interchange of two nuclei as a whole, whilst the nuclear wave function should be antisymmetric on interchange of two protons (or neutrons), either within a nucleus or between two nuclei. If the like nuclei have mass number A then their interchange is equivalent to A interchanges of nucleons. Thus the electronic wave function will be antisymmetric for odd A and symmetric for even A. The symmetry is reflected in the statistical behaviour of an aggregate of such atomic systems. Antisymmetric states conform to the Fermi–Dirac distribution, whilst symmetric states conform to the Bose–Einstein distribution (Martin 1975; Pauling and Wilson 1935); for a diatomic molecule of similar atoms, the intensities of lines in a rotational band spectrum reflect the statistics obeyed and so determine whether the number of nuclear particle components is odd or even (see Burcham 1963). This may sound trivial now, but before the discovery of the neutron it led to rejection of the hypothesis that nuclei were composed of protons and electrons. In particular, for ^{14}N $(14, 7)$ the N_2 molecule obeys Bose–Einstein statistics, although on that theory ^{14}N was thought to consist of 14 protons and 7 electrons—an odd total of fermions.

The symmetry rules impinge on nuclear reactions in a somewhat different way. Consider the system $(\alpha + \alpha)$; its wave function must be symmetric under the interchange of the two α-particles. If these particles can come together to form a state in ^8Be then this state must reflect that symmetry. But ^8Be is more complex than a (2α) structure (e.g. it can also be formed from $^7Li + p$) and each state can be looked upon as a conglomerate of such structures. However, of all the states in ^8Be which exist only those with a certain symmetry can decay into $\alpha + \alpha$. These states must have $J = 0$, 2, 4, etc., as can be shown by a consideration of the wave functions for $\alpha + \alpha$.

Parity

Parity is another aspect of symmetry in quantum mechanics, but one which has no direct correspondence in the classical limit. If the interaction potential of a system has the symmetry property $V(x, y, z) = V(-x, -y, -z) = PV(x, y, z)$, where the operator P has been introduced and defined, and the Schrödinger wave equation has a solution $\psi(x, y, z)$, then $P\psi(x, y, z)$ is also a solution. For a non-degenerate case, the second solution must be a numerical factor times the first, i.e. $P\psi(x, y, z) = \alpha\psi(x, y, z)$. But it is obvious that $P^2 = 1$, corresponding to two inversions, so that $\alpha = \pm 1$. Thus a non-degenerate solution corresponding to this type of potential has either even parity ($\alpha = +1$) or odd parity ($\alpha = -1$).

For a single particle in a central potential there is a one-to-one correspondence between the orbital angular momentum and the parity, which follows from the angular solution of the problem:

$$\text{even } l \equiv \text{even parity}, \qquad \text{odd } l \equiv \text{odd parity}.$$

For a multi-particle assembly like the nucleus, short of a complete solution to the problem, the parity can only be deduced in terms of a model in which the parities of the individual particles can be specified, as in the shell model (Chapter 3). From the fact that the complete wave function can be expanded in terms which are themselves products of the wave functions of the individual particles, it follows that the overall parity is a product of the individual parities if + is assigned to even parity and − to odd parity.

As a constant of the motion, parity plays an important part in nuclear reactions. If in the reaction $X + a \rightarrow W^* \rightarrow Y + b$, where W^* represents a state in the compound nucleus (Chapter 6), the spins and parities of all five particles are known, then conservation of angular momentum and parity severely restrict the orbital angular momentum taken in by a and taken out by b, thus affecting both the angular distribution and the magnitude of the cross-section. As an example assume in the above that X has (spin, parity) $= (0^+)$, a is a proton ($\frac{1}{2}^+$), b is a neutron ($\frac{1}{2}^+$), Y is 2^+, and the compound nucleus W^* is $\frac{3}{2}^-$. Then, conserving angular momentum only gives $l_a = 1$ or 2, $l_b = 0, 1, 2, 3$, or 4, but parity conservation will reduce these possibilities to $l_a = 1$, $l_b = 1$ or 3. For spinless particle and projectile, the accessible states in the compound nucleus are severely restricted, e.g. for $^{12}C + \alpha \rightarrow$ $^{16}O^*$ the excited states must be 0^+, 1^-, 2^+, 3^-, etc.

Electric and magnetic moments

Electric moments

If a charge is distributed over a finite region of space, surrounding a chosen origin, through which an electric field derivable from a potential $U(x, y, z)$ has been imposed from outside, then the interaction energy can be expressed in a Taylor series thus,

$$E = \int U(x, y, z)\rho(x, y, z)\,d\tau = U(0)\int \rho\,d\tau$$

$$+ \left(\frac{\partial U}{\partial x}\right)_0 \int x\rho\,d\tau + \left(\frac{\partial U}{\partial y}\right)_0 \int y\rho\,d\tau + \left(\frac{\partial U}{\partial z}\right)_0 \int z\rho\,d\tau$$

$$+ \frac{1}{2}\left(\frac{\partial^2 U}{\partial x^2}\right)_0 \int x^2\rho\,d\tau + \frac{1}{2}\left(\frac{\partial^2 U}{\partial y^2}\right)_0 \int y^2\rho\,d\tau$$

$$+ \frac{1}{2}\left(\frac{\partial^2 U}{\partial z^2}\right)_0 \int z^2\rho\,d\tau + \left(\frac{\partial^2 U}{\partial x\,\partial y}\right)_0 \int xy\rho\,d\tau$$

$$+ \left(\frac{\partial^2 U}{\partial y\,\partial z}\right)_0 \int yz\rho\,d\tau + \left(\frac{\partial^2 U}{\partial z\,\partial x}\right)_0 \int zx\rho\,d\tau$$

+ higher-order terms.

The first term gives the interaction energy with the field of a point charge at the origin equal to the total charge—the monopole term. The next three terms represent the scalar product of a vector dipole—the dipole moment of the charge distribution—with the vector field at the origin; whilst the next six terms represent the scalar product of two tensors, one the generalized derivative of the vector field and the other the quadrupole moment of the charge distribution. These classical definitions hold for a quantum-mechanical system if we interpret ρ as $e \sum_i \psi_i^*(r)\psi_i(r)$, where the summation i is taken over the charged particles of the system (protons in the nucleus).

The fact that a nucleus has a well-defined parity makes $\rho(r) = \rho(-r)$. Thus every odd moment—dipole, octupole, etc.—will vanish identically. The simplest deformation from spherical results in an ellipsoid of revolution, which, if the z-axis is the symmetry axis, produces a quadrupole-moment tensor which can be specified by a single parameter Q_0 given by $eQ_0 = \int (3z^2 - r^2)\rho\,d\tau$—a definition which conveniently has $Q_0 = 0$ for spherical symmetry. A positive value for Q_0 corresponds to a prolate shape for the nucleus, i.e. elongated along the symmetry axis; whilst negative Q_0 defines an oblate nucleus, i.e. flattened along the symmetry axis. As defined the dimensions of Q are those of an area for which the convenient nuclear unit is the barn (b) (1 barn = 10^{-24} cm^2 = 10^{-28} m^2). Carrying out the integral for

TABLE 1.1

Nucleus	B(MeV)†	Atomic binding (keV)	Ground state J^π	μ(nm)	Q(b)	$\Delta R/R$‡
n	—		$\frac{1}{2}^+$	−1.9135	—	
$^1_1$H	—	0.014	$\frac{1}{2}^+$	+2.79275	—	
$^2_1$H	2.225	0.014	1^+	+0.85735	+0.00282	?
$^4_2$He	28.29	0.079	0^+	0	0	0
$^7_3$Li	39.24	0.190	$\frac{3}{2}^-$	+3.2564	−0.04	?
$^{16}_8$O	127.6	2.0	0^+	0	0	0
$^{35}_{17}$Cl	298.2	12	$\frac{3}{2}^+$	+0.8218	−0.08	−0.1
$^{57}_{26}$Fe	499.9	34	$\frac{1}{2}^-$	+0.0905	0	0
$^{121}_{51}$Sb	1026	180	$\frac{5}{2}^+$	+3.342	−0.53	−0.12
$^{176}_{71}$Lu	1418	370	7^-	+3.180	+8.0	+0.50
$^{181}_{73}$Ta	1452	390	$\frac{7}{2}^+$	+2.35	+3.9	+0.35
$^{209}_{83}$Bi	1640	550	$\frac{9}{2}^-$	+4.080	−0.34	−0.02
$^{235}_{92}$U	1783	690	$\frac{7}{2}^-$	±0.35	±4.1	±0.25

† B is given only to four figures; it is however known to the nearest keV for the light elements and somewhat less precisely for the heavy.

‡ The $\Delta R/R$ have been computed using $R = 1.1A^{\frac{1}{3}}$ fm. A relationship $Q/Q_0 = J(2J - 1)/(J + 1)(2J + 3)$ is used to connect the observed quadrupole moment Q with the intrinsic quadrupole moment Q_o. This projection factor arises because the intrinsic shape precesses about the vector J which itself precesses around the external axis.

ρ = constant within the ellipsoid gives

$$Q_0 = \tfrac{2}{5}Z(a^2 - b^2) \approx \tfrac{4}{5}ZR^2\left(\frac{\Delta R}{R}\right),$$

where a, b are the major and minor semi-axes, R the average nuclear radius, and $\Delta R = a - b$.

Obviously a state $J = 0$ has spherical symmetry and so has zero observable quadrupole moment; less obviously, a state $J = \frac{1}{2}$ also has spherical spatial symmetry and zero observable quadrupole moment. Table 1.1 illustrates these points and also indicates that some nuclei can be strongly deformed with $\Delta R/R \sim 30$ per cent.

Magnetic dipole moment

A similar, but more complicated, expansion of the charge distribution and its interaction with an external magnetic field, taking into account a possible circulation of the charge within the distribution will give a leading term dipole in nature. If in transition to the particle system the intrinsic magnetic moments of the individual particles are taken into account, this dipole term will have two parts: (1) from the circulating currents of the proton distribution which will be proportional to some suitable summation of the orbital angular momenta l_i and (2) from a suitable summation of the intrinsic magnetic moments of both protons and neutrons which will be vectors parallel to the spins s_k. The phrase 'some suitable summation' is used because the resultant magnetic moment of a number of particles depends on how the particles are coupled together; recourse to a model is necessary, as will be apparent in Chapter 3 and Appendix E. At this stage it should be noted that the magnetic dipole moment is a vector, and it is a fundamental result in quantum mechanics that all vector constants of a system are scale factors of one vector, which, for an isolated system, will be the total angular momentum J. Thus vectorially $\mu = gJ$, where g is the suitable scale factor. It is convenient to represent the dipole moment by a scalar, which in the classical limit would represent the magnitude $|\mu|$. Two possibilities exist: $g\{J(J+1)\}^{\frac{1}{2}}$ and $g(M)_{\max} = gJ$. The latter has been adopted.

For the case of a spinless particle of mass m and charge e, $J = L$ and g is the gyromagnetic ratio $e/2m$—a relationship which can readily be verified for a circular orbit, using Ampere's theorem connecting a circulating current with a magnetic shell. This relationship leads to a natural unit $e\hbar/2m$ for the dipole moment. In atoms the mass is that of the electron, defining the Bohr magneton (μ_B); in nuclei it is the mass of the proton defining the nuclear magneton (μ_N), which has the value $5.051 \times 10^{-27}\,\mathrm{J\,T^{-1}}$. The intrinsic magnetic dipole moments of the proton and neutron are respectively $+2.792\,75\,\mu_N$ and $-1.9135\,\mu_N$. A few nuclear magnetic moments are given in Table 1.1.

Problems

1.1. Assuming that the nuclear size parameters are as on p. 10, calculate the quantum number n for which the circular Bohr orbit of K^- around $^{208}_{82}\mathrm{Pb}$ lies in nuclear matter at a density of 10 per cent of the central nuclear density. The mass of the K^- is $494\,\mathrm{MeV}/c^2$.

1.2. By assuming that the neutrons and protons in the nucleus behave as independent Fermi gases, and by neglecting the Coulomb interaction, derive the form of the symmetry term, proportional to $(N - Z)^2$, which appears in the mass formula.

1.3. Verify the statement that the pairing term in the mass equation is sufficiently large to ensure no odd–odd nucleus (above $A = 40$) is less massive than both its even–even neighbours of the same mass number.

1.4*. Calculate values (in MeV/c^2) for γ and ε in the mass equation from the following facts: $^{35}_{18}A$ emits positrons with a maximum energy of 4.95 MeV; $^{135}_{56}Ba$ is the stable isobar of mass number 135.

1.5*. The maximum energy of the positrons from the decay of $^{13}_{7}N$ is 1.24 MeV and there is no subsequent γ-radiation. Deduce a value for the radius of nuclei of mass 13. (The n–p mass difference is 1.29 MeV/c^2.)

1.6. Show that the angular wave function $\sin \theta \exp(i\phi)$, corresponding to $l = 1$, $m_l = 1$, has odd parity, whilst $\sin^2 \theta \exp(i2\phi)$ and $\sin \theta \cos \theta \exp(i\phi)$, corresponding to $l = 2$, $m_l = 2$ and 1, have even parity.

1.7. Deduce the gyromagnetic ratio for the case of a charged particle moving in a circular orbit. Hence derive the natural units for the magnetic moments of electron and proton orbits.

---

* indicates that the problem is taken from, or based on, an Oxford Finals question.

2. Nuclear forces

Introduction

From the last chapter a picture of the nucleus has emerged which is quite different from that of an atom. In the latter, the small massive central nucleus interacts more strongly (~Z times) with any electron than that electron does with any other electron. For the nucleus there is no well-defined centre; the interaction of one nucleon with its neighbour at any instant of time is probably stronger than its interaction with the nucleus as a whole—this latter being a time-averaged effect of all the individual interactions. Thus the problem is a many-body one, without the obvious approximations of the atomic case; in the next chapter models will be illustrated which simplify the many-body aspects. The shell model has been taken over from the atom, but initially with reluctance, since its basic concepts seem to clash with ideas of the nucleus.

In this chapter, the basic nucleon–nucleon interaction is examined in order to answer the questions:

(1) Assuming that the strong interaction between nucleons can be represented by a potential function, what is its basic shape and does it depend on the type of nucleon?

(2) Arguing backwards from a nucleus towards a basic nucleon–nucleon interaction, does this interaction agree with that deduced from basic nucleon–nucleon scattering experiments?

(3) Knowing that the answer to (2) is indefinite, is there a need to introduce new forces in the nucleus which are not present in the basic two-body interaction?

The answer to the third question can be given now. Theorists

29

are not happy with the idea of many-body forces, although there appears to be no very good reason why they should be small compared with the two-body interaction. The subject has not yet reached such a stage of refinement as to need a small 'topping up' with three-body forces. If they are to be introduced at all, it may as well be done in bulk—in which case what of four-body forces, etc? The hope appears to be that averaging many-body forces in a large nucleus produces much the same effect as averaging two-body forces. Note that, in answering this question, addition of a third nucleon will always modify the interaction between two nucleons because of the Pauli principle—but this effect is always allowed for in any attempt to solve the problem of larger nuclei and is not to be looked upon as a new three-body interaction in the sense referred to here.

Inferences from the data presented

In developing the nuclear properties of the previous chapter, a number of facts have emerged which have a direct bearing on the forces which hold the nucleus together. Taken roughly in the order in which they were presented they are:

(1) nuclear sizes are of the order of a few fm;

(2) nuclear density, at least near the centre, is the same for all nuclei;

(3) protons and neutrons have similar density distributions inside the nucleus;

(4) scattering of nuclear projectiles gives the Rutherford cross-section up to a certain angle with a rapid fall-off above this angle;

(5) the basis for a successful analysis of electrostatic energy differences between mirror pairs is that nuclear forces are 'charge symmetric', i.e. $(p, p) \equiv (n, n)$;

(6) the nuclear density falls off over a range of ~ 1 fm at the periphery;

(7) the leading terms in the binding energy, $\alpha A - \beta A^{\frac{2}{3}}$, were arrived at by analogy with a liquid drop;

(8) there is a 'symmetry term' in the mass equation;

(9) the existence of a pairing term, and the coupling to zero angular momentum of every ground state of even A indicate a pairing force;

(10) the α-particle is almost as tightly bound as a massive nucleus;

(11) some heavy nuclei have large quadrupole moments;

(12) even the deuteron has a quadrupole moment;

(13) the proton and neutron have 'anomalous' magnetic moments, i.e. different from those for structureless Dirac particles;

(14) the magnetic moment of the deuteron is close to, but not equal to, the sum of the proton and neutron magnetic moments.

Range and saturation

The terms 'long-range' and 'short-range' have both a general and a specific meaning. In the general sense long and short are relative to the size of the system; van der Waals forces of cohesion of a liquid drop are short-range in the sense that the binding of a given molecule is dominated by the contributions of a few fairly close neighbours rather than by the overwhelmingly greater number of distant molecules—these forces are proportional to $1/r^7$. Likewise, in nuclei, forces are short-range if the contribution of most nucleons to the binding of a given one is less than the contribution from its near neighbours. In the specific sense, forces are said to have a range if the potential function contains an exponential, e.g. Yukawa forces having a potential function $U \propto (1/r)\exp(-kr)$ are said to have a range $1/k$. Thus it will be seen that nuclear forces probably are short-range in both the general and specific senses, though arguments on saturation are concerned really with the general sense.

The nuclear force field extends beyond the nuclear surface by a distance of the order of magnitude of the range of nuclear forces, so (1) and (4) (of the previous section) taken in conjunction can be used to define this range. The estimate obtained is confirmed by (6) since we would, in a simple way, expect the fall-off of nuclear density to occur over the same range.

(2) and (7) in conjunction with the liquid-drop model from which the mass terms were derived do much to define nuclear forces. The molecular force field in a liquid drop has two main components: the attractive short-range force, which can be oversimplified by describing it as an induced-dipole–induced-dipole interaction for which the potential energy is proportional

to r^{-6}; and the repulsive, even shorter-range force which occurs when the electron clouds overlap. By analogy, a similar state of affairs may be expected to occur in nuclei; the short-range attraction would require a range of 1–2 fm, whilst the shorter-range repulsion might have a range of less than 0·5 fm.

Oddly enough, in the historical development of the subject, the concept of a repulsion at short distance was not readily acceptable. Nucleon–nucleon scattering had not yet indicated the need for such a force, which appeared to require a complex structure for what was hoped to be a simple fundamental particle. Instead, recourse was made to the exchange phenomenon (see Appendix C) to provide forces with suitable properties. Such forces were described as: Wigner (W) or ordinary (no exchange of quantum numbers); Majorana (M) or space-exchange (corresponding to exchange of charge and spin); Heisenberg (H) or charge-exchange (but not spin); and Bartlett (B) or spin-exchange (but not charge). All these forces can arise from pion-exchange. From analogy with electron-exchange binding of molecules, a nucleon should be surrounded by a pion cloud. That this is so can be inferred from the magnetic moments of proton and neutron, which for simple Dirac particles should be $1\,\mu_N$ and $0\,\mu_N$; but the pion has a much smaller mass than proton and neutron so the cloud can, and does, make a large contribution to their moments.

Exchange forces will depend upon whether the property being exchanged is symmetric or antisymmetric in the wavefunction representing the relative motion of the two particles. As an example the spin-exchange operator (see Appendix C), $P^\sigma = \frac{1}{2}(1 + 4s_1 \cdot s_2)$, has the value $+1$ for $S = 1$ and -1 for $S = 0$, where $S = s_1 + s_2$, so the spin-exchange force can be made attractive for $S = 1$ and repulsive for $S = 0$, or vice versa.

How do exchange forces lead to saturation? In a heavy nucleus A, the number of nucleon-nucleon interactions is $\frac{1}{2}A(A - 1)$. If the forces are ordinary then all these are increasing the binding of the nucleus, and a complete calculation leads to all nuclei having roughly the same size with a binding energy proportional to A^2 rather than A. However, the operation of the Pauli principle ensures that all nucleon pairs cannot have the same spin symmetry or charge symmetry or spatial symmetry, so that exchange forces will as often be repulsive as attractive. Their overall effect need not be zero, since the strength of the force will depend upon a spatial overlap integral which of course is largest when the spatial wavefunctions of the two particles are the same. The exclusion principle, in its restricted statement, limits the number of particles with the same spatial wavefunction to 4—two protons (spin up, spin down) and two neutrons. Thus Majorana (spatial) exchange forces might be expected to produce strong interactions within groups of four particles and weak effects on averaging over pairs not in the same group. The emergence of four particles is an attractive result because (10) emphasizes the fact that the α-particle is about as strongly bound as any heavy nucleus.

Much effort has gone into producing a recipe for nuclear forces using these four types, but each recipe, whilst successful for explaining the particular fact on which the 'cook' was concentrating, has had its shortcomings, and the fact of a repulsive core (see p. 43) has been finally accepted with relief. This does not mean that exchange forces are unnecessary; it is difficult to see how ordinary forces with repulsion could lead to a saturated α-particle. Indeed, from the range of the repulsive force as required from nucleon–nucleon scattering, it is possible to attach an effective size to each nucleon ($r_{eff} \sim \frac{1}{2} \times$ range, since two nucleons cannot approach closer than this range). The effective volume of the nucleons in a heavy nucleus is then found to be ~ 1 per cent of the nuclear volume. At its face value this indicates that saturation is largely accounted for by exchange forces, though no doubt the repulsive core also plays an important role.

Isospin

In (5) (see p. 30) the point is made that forces are 'charge symmetric', by which it means that the (p, p) and (n, n) interactions are the same, apart from the electromagnetic contribution, but the (p, n) could be different. In the analysis of 'mirror pairs' the numbers of (p, p) and (n, n) interactions are interchanged on going from one nucleus to its mirror pair, but the number of (p, n) interactions is unchanged, so the fact that the energy difference between mirror pairs can be accounted for by the differing Coulomb energies and the proton–neutron mass difference requires that the strong interactions (p, p) and (n, n) be the same, but does not put any restriction on the (p, n) interaction.

A further step to 'charge independence' is made by assuming that all three interactions are the same; this can be checked by analysing nucleon–nucleon scattering or by a more detailed examination of nuclear properties. Charge independence is expressed mathematically using the isospin formalism, which was developed as a convenience before it acquired physical significance. Rather than referring to two distinct types of particle within a nucleus, theorists preferred to employ a nucleon to which an additional two-valued quantum number was assigned and given the value $+\frac{1}{2}$ for the neutron $-\frac{1}{2}$ for the proton. The reason for this choice is that all two-valued quantum numbers have a similar mathematical behaviour, so the values reflect the similarity with $S_z = \pm\frac{1}{2}$ for the spin quantum number.

The name 'isospin' is a shortening of 'isobaric spin', indicating that the two components have the same mass and emphasizing the similarity with spin S. By analogy, there exists a vector τ and a component τ_z of isospin, but the space in which this vector exists is not the same as the space in which ordinary spin exists. Thus, in this space, $\tau_z = +\frac{1}{2}$ represents a neutron and $\tau_z = -\frac{1}{2}$ a proton.† Pursuing the analogy, a vector sum $T = \tau_1 + \tau_2$, corresponding to $S = s_1 + s_2$, can be made for two (and more) nucleons. A nucleon pair can have $T = 1$ giving a function symmetric on exchange of nucleons, or $T = 0$, antisymmetric. If the Pauli principle is extended to make the overall wavefunction antisymmetric with the isospin component included then it can be seen that the same states in the two nucleon configuration are obtained whether or not this formalism is used; since (p, p) and (n, n) are symmetric in isospin their wavefunctions must be antisymmetric in (space) (spin), but the (n, p) combination can be symmetric or antisymmetric in isospin ($T = 1$, $T_z = 0$ represents (p, n) and is symmetric; $T = 0$, $T_z = 0$ also represents (p, n) and is antisymmetric) so the (space) (spin) wavefunctions can also be either.

Thus far, the formalism is merely mathematical, but charge independence can readily be fed into the scheme by asserting

---

† It is unfortunate that particle physicists have later chosen precisely the opposite convention—in principle it does not matter, since the direction up or down in an unspecified space is hardly relevant, but it is confusing.

that nuclear forces depend only on T and not on T_z. This achieves its aim because the selection of T determines the type of (space) (spin) wavefunction and so ensures that we compare like with like when testing the similarity of (p, p), (p, n), and (n, n) interactions. The equality of strong interactions in mirror pairs of nuclei follows from this statement, at least for $|N - Z| = 1$, because these mirror pairs have $T_z = \pm\frac{1}{2}$ and, moreover, have the same value for T, also $\frac{1}{2}$. Extending to the case $|N - Z| = 2$; ^{14}C and ^{14}O, corresponding to interchange of neutrons and protons, have the same value for $T(=1)$, but $T_z = \pm1$. But charge independence further requires that there should exist a state in ^{14}N having $T = 1$, $T_z = 0$ with similar strong-interaction energy as the ground states of ^{14}C and ^{14}O. The same should apply to excited states in ^{14}C and ^{14}O. The state of affairs is shown in Fig. 2.1, where, to emphasize the point, the nuclei have been moved relative to each other through energies to allow for the electrostatic term and the p–n mass difference. An interesting feature is that the ground state of ^{14}N has $T = 0$; there is a

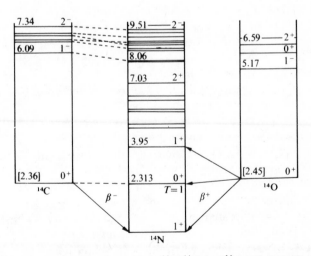

Fig. 2.1. Comparison of energy levels in ^{14}C, ^{14}N, and ^{14}O. Allowance has been made for the Coulomb energy and p–n mass difference. Note that ^{14}C and ^{14}O have $T = 1$ states only, whereas ^{14}N has $T = 1$ states and a greater number of $T = 0$ states. The dotted lines guide the eye to $T = 1$ states in ^{14}N. (Based on Ajzenberg and Selove (1970). *Nucl. Phys.* **A152**, 1.)

general tendency for states of lower T to lie lower which results in the symmetry term in the mass equation (point (8) of p. 30). The method used to derive it implied charge independence since it treated neutrons and protons in a similar way, which again is implied in point (3).

Thus another quantum number has been introduced into the description of a nuclear state but, as the shift of energy in Fig. 2.1 has perhaps indicated, it is only an approximate one, since the electromagnetic term, whilst splitting the isospin sub-states by a large energy (several MeV), is also of a form which mixes states with different T to some (small) extent. The charge on a nucleon can be written as $(\frac{1}{2} - \tau_z)e$ so the Coulomb interaction will contain products of such terms. The bracket contains an isoscalar and the third component of an isovector; the former does not mix isospin states, but the latter operates upon a pure isospin state to give a mixture of isospins. With such a term in the Hamiltonian, the eigenstates cannot be pure isospin states.

This quantum number plays a significant role in nuclear reactions; e.g. the reaction $^{16}O + d \rightarrow \, ^{14}N + \alpha$ leaves ^{14}N in excited states which are characterized by $T = 0$, since ^{16}O (ground state), d, and α all have $T = 0$. Since the state at 2.31 MeV belongs to the $T = 1$ triplet which includes the ground states of ^{14}C and ^{14}O, it can be populated in this reaction only by way of the small impurity in the wavefunction of the $T = 0$ component. This selectivity in (d, α) and similar reactions had been noticed experimentally and led to the conclusion that T was indeed quite a good quantum number.

Finally, to return to the mass equation, the pairing term (point (9)) appears to refute charge independence in favour of charge symmetry, since the term increases the binding for (p, p) and (n, n) pairs but decreases it for (p, n). However, we are failing to compare like with like; the proton pair, or neutron pair, have similar quantum numbers — this will be more obvious after reading about the shell model in the next chapter—whilst in a large nucleus the (p, n) pair are in dissimilar orbits. However, in the very light nuclei when the (p, p), (n, n), and (p, n) interactions all refer to nucleons in the same orbits (shells), it is interesting to note that the odd-odd nucleus is stable on four occasions. Thus the pairing term can be shown to be consistent with charge independence.

The deuteron and the virtual deuteron

Since the deuteron presents a simple two-body problem, its study was hoped to reveal the properties of the nucleon–nucleon interaction. In hindsight it has proved to be somewhat of a disappointment; the solution for any simple short-range potential function which might approximate to the nucleon–nucleon interaction, e.g. the square well having potential $-V_0$ from $r = 0$ to b and zero for $r > b$ (see Fig. 2.2) indicates that, in the loosely bound deuteron, the nucleons are for most of the time outside the effective range of the force — farther apart than b in the square well quoted above. To pursue this example it turns out that the wavefunction can be very similar for a large range of variation of the square-well parameters, provided that such variation keeps $V_0 b^2$ approximately constant. The value of the constant is $\sim 10^{-24}$ MeV cm^2 i.e. 1 MeV barn corresponding to something like $V_0 \sim 50$ MeV, $b \sim 1.4$ fm. This condition follows from that fact that $Kb \sim \frac{1}{2}\pi$, where K is the wavenumber inside

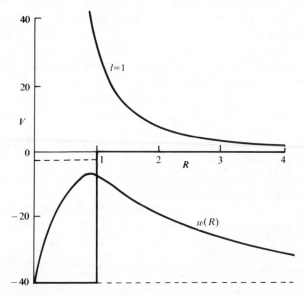

Fig. 2.2. Typical square well for the deuteron, depth 40 MeV, size 1.6 fm (the scale factor for the abscissa). Also shown is the centrifugal barrier for $l = 1$, and $u(r) = rF(r)$, where $F(r)$ is the radial wavefunction of the ground state of the deuteron.

the potential, for successful matching of the internal and external wavefunctions to produce a stationary state. For a more realistic potential without discontinuities, a similar condition can be deduced in the form of a phase integral in the region of positive kinetic energy. It need hardly be added that the analysis of the deuteron sheds no light on the existence of a repulsive core. Also, because of the fact that the deuteron size stems mainly from that part of the wavefunction corresponding to negative kinetic energy and having a (slow) exponential fall-off, feeding into the problem the (mean) size from electron scattering does not serve to pin down the force range with any precision.

An interesting result arises, however. The lowest state is undoubtedly an S-state. That this state is bound only by 2.2 MeV, in a well of depth ~50 MeV, makes no other bound state possible; the next S-state with a node in the wavefunction requires a wavenumber of approximately 3 times that of the present state, which is clearly impossible, and the lowest $l = 1$ state is made unbound by the 'centrifugal potential'. Classically, a circular rotor with angular momentum $m\omega r^2$ has kinetic energy $= \frac{1}{2}m\omega^2 r^2$, which becomes $l(l + 1)\hbar^2/2mr^2$ on quantizing the angular momentum to $\{l(l + 1)\}^{\frac{1}{2}}\hbar$. A term of this form automatically arises in the radial Schrödinger equation — it is shown in Fig. 2.2 for $l = 1$. Numerically, to bind a state having $l = 1$ would require a well-depth of about 200 MeV for the same range. It is high unlikely that a mixture of exchange forces could be produced capable of such large variation with l of the potential.

However, it is known from low-energy neutron scattering by protons that there exists a virtual state in the deuteron unbound by ~30 keV. The detailed analysis of scattering is beyond the scope of this discussion. Suffice it to say that from measurements of the differential cross-section it is possible to deduce the presence of resonances (or virtual states) and to determine the orbital angular momentum l taken in by the neutron in forming this virtual state. The present experiments were performed at thermal energies and could only detect such a resonance for $l = 0$ (see later in chapter). Thus the virtual state is also an S-state.

For the simple potential $V(r)$ there should be spin-degeneracy giving two S-state wavefunctions, namely, $J = 1$, $^3S_1$, and $J = 0$, $^1S_0$. That they are not degenerate is proof of the spin-dependence of nuclear forces and therefore of exchange — but the analysis does not lead to a recipe for the mixture. The ground state is obviously spatially symmetric and spin symmetric; in isospin it must be the antisymmetric $T = 0$ state. The virtual state is spatially symmetric and spin antisymmetric; in isospin it must be the symmetric $T = 1$, $T_z = 0$ state. From the operator forms of the different exchanges (see Appendix C) it is seen that the Majorana term should be the same for both states, but that both Heisenberg and Bartlett terms should change. In the absence of any other states of the deuteron system, no evidence is available on the Majorana term; the effect of the other two terms would appear to be small, but the overall picture is far more complex than presented here so no quantitative conclusions are drawn.

So far, a simple theory has assigned the characteristics $^3S_1$, $J = 1$, parity even—denoted by 1^+—to the ground state of the deuteron. An S wavefunction is necessarily symmetric in space, so the deuteron should have no quadrupole moment. But, as stated in point (12) (see p. 30), the deuteron does have a (small) quadrupole moment. There can be no doubt that 1^+ is correct since it is soundly based experimentally. The only other possible wavefunction having 1^+ is the $^3D_1$-state, i.e. the coupling of spin 1 to $L = 2$ ($L = 1$ has odd parity) to give $J = 1$. This wavefunction cannot by itself form the ground state since it has already been shown that no state possessing orbital angular momentum can be bound by any reasonable potential, but in any case it would give too large a quadrupole moment. The remedy is to mix a small amount (~4 per cent probability) of this wavefunction with the basic $^3S_1$ wavefunction. But this mixture no longer has a well-defined value for L; L is certainly a constant of motion for any central force, and so a non-central term must be included in the potential. Such an interaction is already present. The two nucleons have magnetic dipole moments giving a dipole–dipole interaction of the form

$$V_{dd} \propto r^{-3}\{3r^{-2}(\boldsymbol{\mu}_1 \cdot \mathbf{r})(\boldsymbol{\mu}_2 \cdot \mathbf{r}) - (\boldsymbol{\mu}_1 \cdot \boldsymbol{\mu}_2)\}, \quad \text{i.e.}$$
$$\propto r^{-3}\{3r^{-2}(\boldsymbol{s}_1 \cdot \boldsymbol{r})(\boldsymbol{s}_2 \cdot \boldsymbol{r}) - (\boldsymbol{s}_1 \cdot \boldsymbol{s}_2)\} = r^{-3}S_{12}.$$

This magnetic interaction is, however, quite negligible— something of the order of a few hundred eV. It is therefore necessary to introduce a strong interaction of the form $V(r)\, S_{12}$, where $V(r)$, some radial function with a range, replaces the electromagnetic function r^{-3}. To illustrate that it mixes l-values, take for simplicity the term in S_{12} arising from the z-components of both spins (there are nine terms in the product $(s_1 \cdot r)(s_2 \cdot r)$). It is

$$s_{1z}s_{2z}\left(\frac{3z^2}{r^2} - 1\right) = s_{1z}s_{2z}(3\cos^2\theta - 1) \propto s_{1z}s_{2z}Y_{20}(\theta\phi),$$

where $Y_{lm}(\theta\phi)$ is a spherical harmonic, and, when found in the wavefunction of a particle, indicates that the particle has orbital angular momentum l and component along the z-axis m. Now allow this component to act on the $M = 1$ substate of the $^3S_1$ state which is a constant in angular dependence, since $l = 0$, and is represented by $\uparrow\uparrow$ in spin, i.e. particle 1 has spin up ($m_1 = +\frac{1}{2}$) and particle 2 has spin up ($m_2 = +\frac{1}{2}$) giving $M = 1$. It therefore produces a state of spatial wavefunction $Y_{20}(\theta\phi)$ which is a combination of a spatial D-wave and an $S = 1$ spin wavefunction having $M = 1$. From examination of this small part of the total interaction of S_{12} it is not possible to assert that this form represents just a $^3D_1$ state since it would also occur in $^3D_2$ and $^3D_3$. It is necessary to examine the whole expression for S_{12} to eliminate these, but from general grounds we can do so, since the system must have a well-defined J. Thus this operator acting on a pure $^3S_1$ state will produce a D-state admixture. For a wavefunction to be an eigenfunction of the potential $V_0 + VS_{12}$ then it must be a suitable mixture of $^3S_1$ and $^3D_1$ to begin with.

The S_{12} term is obviously different for the parallel spin orientations of Fig. 2.3. Since the deuteron is prolate, the

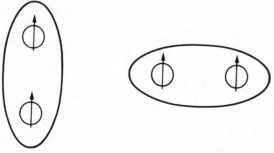

Fig. 2.3. Schematic representation of the spin–spin interaction in the deuteron. The spins are shown directed along the symmetry axis.

left-hand configuration must lead to an attractive force, giving a negative sign to the potential term above. Note that the large quadrupole moments of heavy nuclei (point (11)) result from correlated motions of a number of nucleons and can arise with central forces between nucleons.

This admixture of $^3D_1$ in the wavefunction is also required in accounting for the magnetic dipole moment of the deuteron. As stated in point (14), the magnetic moment is almost equal to the sum $\mu_p + \mu_n$; for a pure $^3S_1$-state it should be exactly equal. Classically this statement is obvious; quantum mechanically it is much less obvious. A calculation along the lines of the Landé approach to hyperfine structure in atoms (see e.g. Kuhn 1961 and Appendix E) is needed to produce this result and also to determine the magnetic moment of the $^3D_1$-state. Adding ~four per cent admixture of $^3D_1$ moves the magnetic moment to the right value. At this stage a note of caution should be injected. As stated in point (13) (p. 31) the magnetic moments of the free proton and neutron are anomalous. This can be ascribed to meson fields surrounding these particles; the proton looks like $(n + \pi^+)$ for part of its time, and the neutron like $(p + \pi^-)$. Since pion masses are considerably less than nucleon masses they make a large contribution to the magnetic moment. What happens when nucleons are in proximity and mesons can then interchange between them? One can imagine the meson cloud spreading out and, in the case of a heavy nucleus, perhaps pervading the whole nucleus. This casts doubt on the determination of nuclear magnetic moments simply by vector addition over individual nucleons—but in the case of the deuteron the two nucleons are probably sufficiently far apart for this simple addition to be correct. Even so, the simple theory of magnetic moments given in Appendix E neglects relativistic effects and could be in error by a few per cent. Fitting the quadrupole moment of the deuteron depends upon the precise recipe for the nuclear force including its tensor component with the result that both moments pin down the D-state admixture to the range ~4–7 per cent in intensity.

From the fact that the admixture of $^3D_1$ is small, it would be hasty to conclude that the non-central part of the potential is also small compared to the central terms. The 'centrifugal potential' for the D-state for such a small system serves to reduce its effectiveness to such an extent that it becomes important to keep the range of the tensor force at least as large as that of the rest of the potential, in order to keep its magnitude down to an acceptable value. Both the exchange central force and the tensor force can be derived from pion exchange. The fact that the pion has spin 0 but negative parity means that it must exchange with $l = 1$ in order to preserve nucleon parities. This transfer of orbital angular momentum can cause the spin-flip necessary in some exchange terms and can also produce the tensor term. The exchange of a pion pair (0^+) results in predominantly S-wave transfer and an ordinary force (the combination of two pions gives $T = 2$ or $T = 0$ symmetric and $T = 1$ antisymmetric; these must combine with a symmetric and an antisymmetric space function respectively. Thus $T = 0$, $S = 0$, $L = 0$ is the only possibility for exchange). Since it is also repulsive it is believed to be (partly) responsible for the repulsive core.

Nucleon–nucleon scattering

The (n, p) scattering system at thermal neutron energies has been cited (p. 38) as a source of information on the $^1S_0$ virtual state of the deuteron. Scattering at higher energies with this system and the (p, p) system gives much more information provided that the energy is high enough. Wavemechanically, the projectile is represented by a plane-wave passing the target, placed at the centre of the coordinate system. Such a plane-wave can be analysed into an infinite series of spherical waves converging on and diverging from the coordinate centre and characterized by the value of the orbital angular momentum l. This series is the quantum-mechanical equivalent of the classical sub-division of the wavefront into zones characterized by the impact parameter of the collision; if this latter parameter is p then $pmv = \hbar\{l(l+1)\}^{\frac{1}{2}}$ is the correspondence, where v is the velocity of the projectile of reduced mass m. It is interesting to put numbers into this equation, using the nucleon–nucleon system to define m: for $l = 1$, 20 MeV incident neutrons in the laboratory system will correspond to a classical impact parameter of 3 fm. The $l = 1$ partial wave of the plane wave corresponding to 20 MeV neutrons will therefore be unaffected by the collision if the interaction is weak at a distance of 3 fm (the corresponding distance at $E_n = 1$ MeV is 13 fm). Thus there is a need for higher and higher projectile energies in order to determine the interaction for higher l-values. Although thermal neutrons were sufficient for the S-wave interaction, they could give no information on the interactions at higher l-values for which laboratory energies ~ 30 MeV are necessary. At $E_n \sim 300$ MeV, the impact parameter is down to 0·7 fm for $l = 1$, so information can be obtained on the potential at close approach.

Note that this same equation with $\{l(l+1)\}^{\frac{1}{2}} = 1$ is also an expression of the uncertainty principle; if the wavelength in the plane wave is λ, then this distance is also the measure of the uncertainty with which a particle can be located and is therefore a limit to which the system can be investigated. Thus for the S-wave interaction an energy of ~ 300 MeV is also required to probe the repulsive core. It is from experiments performed at these energies that the need arose for a repulsive core in the nucleon–nucleon interaction.

Another result can be discerned from a qualitative examination of neutron–proton scattering (see Fig. 2.4). The scattering in the centre-of-mass system rises sharply toward 0° and 180° at high bombarding energies. Peaking at 0° is to be expected; scattering through large angles requires a head-on collision (or nearly so), whilst small-angle scattering can occur when the neutron passes the proton at a distance and so experiences the extreme tail of the potential. Many more neutrons will 'miss' the proton than

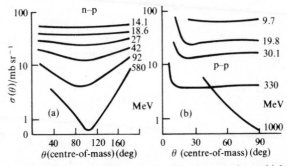

Fig. 2.4. (a) n–p scattering at high energies. (b) p–p scattering at high energies; note, because of similarity of striking and struck particle, the angular distribution must be symmetrical about 90° (centre-of-mass). (Based on Hess (1958). *Rev. mod. Phys.* **30,** 368.)

'hit' it. This näive picture conforms with analysis provided the potential is simply $V(r)$, i.e. an ordinary, central potential. The rise at $180°$ is proof of charge-exchange, corresponding to a small-angle deflection coupled with exchange of charge; the detected neutron, which was originally the proton, moves in the centre of mass in the opposite direction. The type of exchange has been described as charge (or Heisenberg); it could equally have been charge-spin (or Majorana), since the measurements shown refer to unpolarized target and projectile and therefore are averaged over spins. As has been stated previously a large Majorana term is already required.

Another point arising from nucleon–nucleon scattering is that the charge-independence hypothesis appears to be well founded. This is not a result which can be arrived at qualitatively; indeed the qualitative conclusion (see Fig. 2.4) might well be the opposite. The (p, p) system is, of course, affected by Coulomb forces, and there is an added complication that the two particles are identical, giving rise to interference terms when the wavefunction is correctly made antisymmetric. Even when these differences are taken into account, the comparison of (p, p) and (n, p) is still not like with like: the (p, p) system must have $T = 1$ whilst the (n, p) has equal admixtures of $T = 1$ and $T = 0$. The analysis is lengthy and complicated, but the final result appears to be that the $T = 1$ component of (n, p) behaves similarly to the (p, p) when allowance is made for Coulomb and symmetry effects. Note that the strong interaction due to exchange of pions cannot be strictly charge-independent since the (n, n) and (p, p) interactions arise only from exchange of π^0 but the (n, p) interaction, even in the $T = 1$ combination, arises from exchange of all three types of pion, π^0 and $\pi^\pm$. However the masses of the charged pions are slightly different from the mass of π^0 ($M_{\pi^0} = 135 \, \text{MeV}/c^2$, $M_{\pi^\pm} = 139.6 \, \text{MeV}/c^2$) due to the differences in electromagnetic structure, so the ranges must be correspondingly different. The differences are quite small and, as far as nuclear structure is concerned, far less important than the mixing of states of different T due to the Coulomb potential. Also note that a particle physicist, as distinct from a nuclear physicist, would not make the above distinction between the two types of electromagnetic interaction and therefore would interpret the strong interaction as isospin conserving.

One final piece of information has arisen from nucleon–nucleon scattering with polarized beam and target. A spin–orbit coupling term $(\boldsymbol{L} \cdot \boldsymbol{S})$ appears to be necessary. Again it has a magnetic analogue, but again it is far too weak. In introducing a strong interaction of this type, the radial function has been given a short range since the term seems to become more effective at higher energies.

Summary

Nuclear forces have been shown to have:
  (1) a short range, of order 1.5 fm;
  (2) a considerable exchange component, largely of the Majorana type;
  (3) a repulsive core, of range ∼ 0.5 fm;
  (4) a tensor component;
  (5) a velocity-dependent spin–orbit interaction.
The greater part of these requirements can be met by a potential derived from exchange of a pion, giving a range of 1.4 fm and accounting for the exchange and tensor components.

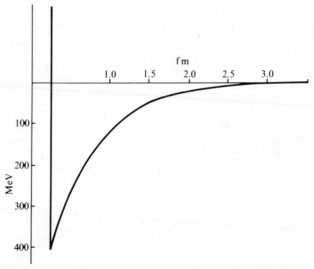

Fig. 2.5. An indication of the nucleon–nucleon potential, with hard-core and pion-exchange forces. Note that, in detail, the potential depends upon the combination of T, L, S, J.

The repulsive core can be accounted for by simultaneous exchange of two (or more) pions or by exchange of heavier mesons. Among these heavier mesons must be included a vector meson $(J^\pi = 1^-)$ in order to account for the spin-orbit term.

A rough qualitative picture of the nuclear potential is given in Fig. 2.5. Notice that in detail it should be different for different combinations of the quantum numbers L, S, and T.

Problems

2.1. Consider the Schrödinger Equation for the spherically symmetric square-well potential: $V = -V_0$ for $r = 0$ to b and $V = 0$ for $r > b$. Show (1) that a term $l(l+1)\hbar^2/2mr^2$ arises in the differential equation for the radial component of the wavefunction in the expression for the potential and (2) that for $l = 0$ solutions are of the form $\frac{1}{r}u(r)$ where $u(r)$ are solutions of the one-dimensional Schrödinger equation for a well of the same shape, but with $V = \infty$ for $r < 0$. (In the 3D case r is positive only by definition of the coordinate system; putting $V = \infty$ for negative r confines the 1D solution to positive r only.)

2.2. If the n–p potential is of the form $V_1(r) + V_2(r)\mathbf{s}_1 \cdot \mathbf{s}_2$, show that J^2, parity, L^2, and S^2 are constants of the motion.

2.3. Determine the magnetic moment of the deuteron in the $^3D_1$-state and show that a few per cent admixture of this state into the $^3S_1$-state gives the measured magnetic moment (see Appendix E).

2.4. Show that the formal introduction of isospin together with a generalization of the Pauli principle does not affect the number of states available to two nucleons.

2.5. The Serber potential is of the form $V(r)(1 + P^s)$ (see Appendix C). Show that it gives no force in odd spatial states.

2.6*. The ground states of $^{10}_4$Be and $^{10}_6$C lie respectively 0.56 MeV and 3.58 MeV above the 3^+ ground state of $^{10}_5$B. Estimate the energy and spin of one excited state of $^{10}_5$B. Comment upon the possibility of exciting this state by deuteron inelastic scattering.

2.7*. Discuss what facts you can deduce about the interaction between two nucleons from the following observations:

(a) in the elastic scattering of neutrons by protons the angular distribution of the scattered neutrons is nearly isotropic in the centre-of-mass system for incident energies below 10 MeV;

(b) when the incident energy is raised above 100 MeV, sharp peaks develop at 0° and 180°;

(c) the binding energy per nucleon in medium-weight nuclei is approximately constant and independent of mass number;

(d) the cross-sections for scattering of slow neutrons by ortho- and para-hydrogen differ by a large factor.

3. Nuclear models

In the previous chapters, simple nuclear models have been introduced, namely, the liquid-drop and Fermi-gas models, which are in effect classical and quantum aspects of the same system. This chapter will discuss three models: the shell model, the collective model, and the optical model. The first is an extension of the Fermi-gas model, whilst the last is a dynamical model concerned with the nucleus at highish excitation, where the restriction of the Pauli principle is decreasing in importance and the system is undergoing transition to something like a liquid drop. In the collective model the restriction to spherical symmetry is removed and the nucleons correlate their motion in the well in such a way as to create the non-spherical well in which they are moving.

The shell model

The atomic shell model has had a great deal of success in accounting for the construction of atoms and the chemical properties of the elements. A brief description is therefore appropriate.

Solution of the Schrödinger equation for the simple system of one electron bound by a central nucleus of charge Z units results in energy levels characterized chiefly by the principal quantum number n. Neglecting the fine structure, each level is degenerate having orbital angular momenta $l = 0, 1, \ldots, (n-1)$; since each level of angular momentum has $(2l+1)$ spatial substates, and in addition there is a spin degeneracy of 2, the total degeneracy is $2n^2$. Also the scale factor of the radial wavefunction is largely determined by n. In filling up such an atom with its Z electrons, account must be taken of the Pauli principle. Thus the first two electrons go into the lowest energy orbits ($n = 1$) and couple to zero angular momentum; their combined wavefunction is therefore spherically symmetric. The next eight electrons go into orbits with $n = 2$, which are less tightly bound. They are therefore on average much further from the nucleus than the other two, and in addition six of the eight have one unit each of angular momentum which tends to keep them away from the centre. Thus these electrons rarely penetrate the spherical cloud which represents the other two and so move around a central nucleus of effective charge $(Z - 2)$. This electron group corresponds to a distribution of charge in the form of a broad spherical shell (in the classical sense), overlapping very little with the shell below. So we build up a complex atom shell upon shell, rather like an onion. The picture is oversimplified and leads to incorrect numbers of electrons in shells ($2n^2$). The reason is that the interpenetration of shells does take place. Because of the centrifugal effect sub-shells of lower l will penetrate the shell below to a greater extent than those of higher l (and the same value of n), so removing the degeneracy enough to move sub-shells effectively from one major energy shell to another. A complete sub-shell, however, is spherically symmetric, so the overall picture is much the same but the numbers of electrons in major shells is now different. This has a profound effect on the chemical properties of the elements.

In trying to adapt this approach to the nucleus, a number of difficulties arise. In both cases the potential well for the next particle must be built up from the interactions of particles already present. In the atom there is initially a strong attraction to the centre, which carries virtually all the mass and provides a basis for building up layers, in addition ensuring that the layers are physically well separated. A central force is therefore dominant at all stages, since a uniformly charged spherical shell behaves, in the region outside it, as though the charge is concentrated at its centre. From Chapter 1, however, it is known that there is nothing special about the centre of the nucleus except that it is in a region where the density is constant and where, from symmetry, the average force on a nucleon is zero. Thus the simplest form of potential which might approximate to the average nucleon–nucleus potential is a square well, whilst the

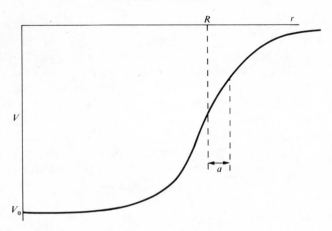

Fig. 3.1. Saxon–Woods (or Fermi) approximation to the nucleon–nucleus potential.

next approximation is to make the potential well follow the shape of the density (see Fig. 3.1) but extending further out by 1 fm or so, to account for the range of the nucleon–nucleon force. With such wells the wavefunctions for the single-particle eigenstates overlap strongly. The centrifugal potential pushes the wavefunction away from the centre for $l \neq 0$, but for $l < K R_{\mathrm{N}}$, where K is the wavenumber of the nucleon inside the nucleus of radius R_{N}, the probability distribution of the nucleon wavefunction is well spread inside the nucleus. Some neutron wavefunctions are given in Fig. 3.2 for a square-well potential for a medium weight nucleus ($R_{\mathrm{N}} \sim 5.5$ fm and $K R_{\mathrm{N}} \sim 8$—this latter value depends of course on the depth of the potential well and the binding energy of the neutron in these states). The states illustrated are obviously the first $l = 4$ and $l = 5$ states since the matching of the interior and exterior parts of the wavefunction requires that the interior part be decreasing in magnitude, to fit the simple exponential exterior part; by the same token the $l = 2$ state is the second of its kind and the $l = 0$ state the third of its kind. The lowest $l = 0$ state, which will have a monotonically decreasing wavefunction (and a much lower value of K to push the first zero out beyond the nuclear radius and therefore to infinity), is therefore seen to overlap strongly with these states, although it represents neutrons in the lowest lying shell. 'Neutrons' is used in

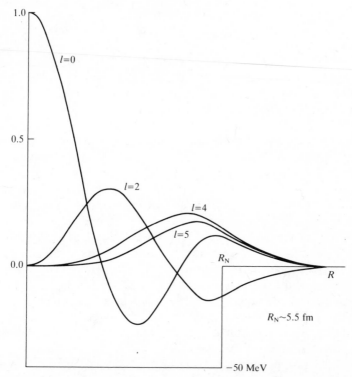

Fig. 3.2. Unnormalized radial wave functions for neutrons in a finite square well potential representing a medium weight nucleus. These uppermost bound levels have $l = 0$, 5, 2 and 4 in order of increasing binding with $KR_N \sim 8$ for $l = 0$ and ~ 7.5 for $l = 4$. Note that the normalizing integrals have as integrand $r^2 |\psi|^2$ so the large central peak for $l = 0$ does not account for a correspondingly large fraction of the complete particle—the integrals are dominated by the peaks in the vicinity of the surface.

the plural here since the simple central potential does not differentiate between spin couplings.

In addition to greater mixing of wavefunctions enhancing the importance of collisions, there is also the fact that the nucleon–nucleon interaction is short-range so that its contribution to the average potential is much less than its contribution to collision. It would seem improbable that the picture of single-particle orbits in an average central potential should approximate to reality. But so far no account has been taken of the Pauli principle. If one particle gains energy in a collision, then the other particle must lose it. In the ground state of a nucleus all the lowest-lying single-particle states will be occupied, so no particle can lose energy. The same two states must be occupied before and after the collision, so effectively there has been no collision. The mean free path of a bound nucleon in the ground state of a nucleus is effectively infinite, and the concept of particle orbits is meaningful.

Having established the picture of nucleons in well-defined states it is now appropriate to consider whether these states group into shells, characterized by having similar energies but not necessarily similar spatial configurations as in atoms. In the case of atoms, the periodic table preceded its explanation in terms of the shell model, so for nuclei statistical analyses of the properties of nuclear ground states revealed the 'magic numbers' which are the nuclear equivalent of the location of noble gases in the periodic table. We take a nuclear property such as binding energy B and plot it against A, or, better still, make a three-dimensional plot against N and Z. Thus in Fig. 1.4 (p. 16) the measured values of B/A differ from the smooth curve predicted by the mass equation in the direction of increased binding in the region of certain values of N or Z. Statistical analysis reveals the following facts:

1. A plot of binding energy of the last nucleon against N (or Z) reveals discontinuities at certain values of N and Z. (The neutron binding energy may conveniently be determined from the γ-spectrum following capture of thermal neutrons.)

2. Similar discontinuities in the binding of α-particles occur at $N = 82$ and 126.

3. Certain nuclei appear to be naturally more abundant than their neighbours in the plot of stable species. Certain elements have more stable isotopes than their neighbours, i.e. a greater variation of N for a given Z, and correspondingly there are values of N for which a greater variation of Z is permitted.

4. The excitation energies of the first excited states (2^+) of even–even nuclei are at maxima for certain values of N or Z, falling to broad minima between these values. In the broad minima the lowest states $(0^+, 2^+, 4^+)$ form a sequence in which the energy spacings $(0, 2)$ and $(2, 4)$ are in the ratio $6:14$; near the maxima these states tend to be equally spaced.

5. Quadrupole moments of odd-A nuclei are large when

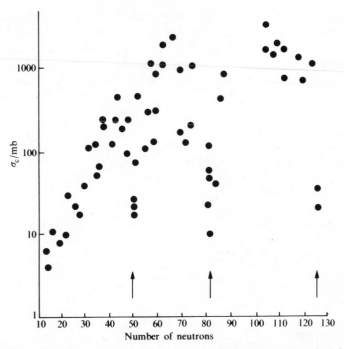

Fig. 3.3. Effective capture cross-sections for reactor neutrons plotted against neutron number of target nucleus. Detailed explanation of this curve is complicated, but the influence of shell closure is apparent. (Based on Codd *et al.* (1956). *Prog. nucl. energy* **1**, 296.)

neighbouring even–even nuclei have low-lying first excited states and are near zero when those states lie high.

6. Nuclear reaction cross-sections reveal periodic properties, e.g. thermal-neutron capture cross-sections (as shown in Fig. 3.3).

7. The parities of nuclear ground states of odd-A nuclei exhibit a systematic behaviour correlated with the above 'magic' effects.

The nett result of such analyses is that nuclei containing 2, 8, 20, (28), 50, 82, 126 neutrons or protons (apart from 126) are particularly stable (cf. the stability of noble gases). By analogy, shell closure at these neutron or proton numbers is therefore expected. The single-particle energy levels of nuclei should reflect such closure by a larger than usual energy spacing at appropriate places in the level diagram. Simply altering the shape of the central potential through permitted variations is unable to produce energy gaps at these numbers (see Fig. 3.4, where the orderings of levels for four types of well is shown to be unsuitable), but the addition to the potential of a spin-orbit term

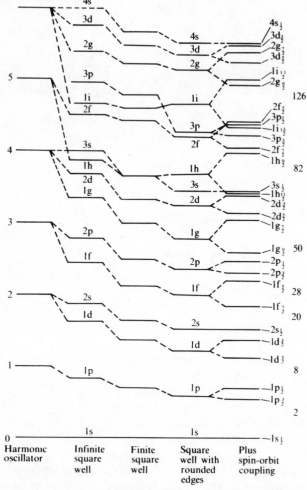

Fig. 3.4. Locations of energy levels in four types of well, followed by the effect of a spin-orbit term on the fourth type. Note that the spin-orbit term is necessary to locate the magic numbers, as shown. (Based on Feld (1953). *A. Rev. nucl. Sci.* **2**, 239.)

($\equiv V(r)\boldsymbol{l} \cdot \boldsymbol{s}$ for a given nucleon) achieves success. This suggestion was made independently by Mayer and by Haxel, Jensen, and Suess in 1949. Such a term splits the otherwise degenerate sub-shells $j = l + \frac{1}{2}$ and $j = l - \frac{1}{2}$ obtained by combining the spin and orbital angular momenta of a nucleon.

Since

$$\boldsymbol{s} \cdot \boldsymbol{l} = \tfrac{1}{2}(j^2 - l^2 - s^2)$$

then

$$(\boldsymbol{s} \cdot \boldsymbol{l}) = \tfrac{1}{2}\{j(j+1) - l(l+1) - s(s+1)\},$$

giving the value $+\frac{1}{2}l$ for $j = l + \frac{1}{2}$ and $-\frac{1}{2}(l+1)$ for $j = l - \frac{1}{2}$, except for $l = 0$, when there is only one value of j, namely $j = \frac{1}{2}$ and $\boldsymbol{s} \cdot \boldsymbol{l} = 0$. Thus the splitting is proportional to $(2l + 1)$; its increase with l is important for the success of the modification since it serves to throw down, from one shell to another, one of the components of the sub-shell with highest l. By making the sign of the spin-orbit coupling opposite to that arising from a magnetic term (as in atoms) the sub-shell corresponding to $j = (l + \frac{1}{2})$ is correctly thrown down. The order of the filling of neutron shells is then given by the sequence

$$(1s_{\frac{1}{2}})^2 \mid (1p_{\frac{3}{2}})^4(1p_{\frac{1}{2}})^2 \mid (1d_{\frac{5}{2}})^6(2s_{\frac{1}{2}})^2(1d_{\frac{3}{2}})^4 \mid (1f_{\frac{7}{2}})^8 \mid (2p_{\frac{3}{2}})^4$$
$$2 \qquad\qquad 8 \qquad\qquad\qquad 20 \qquad\qquad 28$$

$$(1f_{\frac{5}{2}})^6(2p_{\frac{1}{2}})^2(1g_{\frac{9}{2}})^{10} \mid (2d_{\frac{5}{2}})^6(1g_{\frac{7}{2}})^8(3s_{\frac{1}{2}})^2(2d_{\frac{3}{2}})^4(1h_{\frac{11}{2}})^{12} \mid (2f_{\frac{7}{2}})^8$$
$$50 \qquad\qquad\qquad\qquad 82$$

$$(1h_{\frac{9}{2}})^{10}(3p_{\frac{3}{2}})^4(2f_{\frac{5}{2}})^6(3p_{\frac{1}{2}})^2(1i_{\frac{13}{2}})^{14} \mid (2g_{\frac{9}{2}})^{10}(3d_{\frac{5}{2}})^6(1i_{\frac{11}{2}})^{12}(2g_{\frac{7}{2}})^8$$
$$126$$

This sequence can be varied slightly by small changes in the spin-orbit potential without affecting the magic numbers.

The notation is slightly different from the atomic convention. The l_j convention is the same but the number preceding does not have the significance of n in the hydrogen-like spectra; it merely numbers the states of given l_j in the order in which they occur (it can be connected with the number of nodes in the radial wavefunction). The superscript is the number of nucleons which the sub-shell can accommodate $(2j + 1)$. For protons, because of the influence of the Coulomb potential, the ordering is slightly different; it is the same up to 50 when the next shell looks like

$$(1g_{\frac{7}{2}})^8(2d_{\frac{5}{2}})^6(1h_{\frac{11}{2}})^{12}(2d_{\frac{3}{2}})^4(3s_{\frac{1}{2}})^2$$

and above 82 the sequence is

$$(1h_{\frac{9}{2}})^{10}(2f_{\frac{7}{2}})^8(3p_{\frac{1}{2}})^2 \ldots$$

These sequences, together with the pairing term, which is taken to couple pairs of like nucleons to zero angular momentum whenever possible, account for all the parities of the ground states of stable and nearly stable nuclei and, with a few exceptions, their spins for all even-A nuclei and up to $A \sim 100$ for odd-A nuclei. Two notable exceptions are ^{19}F and ^{23}Na which are $\frac{1}{2}^+$ and $\frac{3}{2}^+$ whereas the sequence of protons predicts $d_{\frac{5}{2}}$, i.e. $\frac{5}{2}^+$ for both. However, the $s_{\frac{1}{2}}$, $d_{\frac{3}{2}}$-states lie in the same shell and it was also discovered that a $\frac{5}{2}^+$ excited state lies low in each nucleus, so it was felt that 'residual forces' (i.e. forces left over after accounting for the well by averaging the nucleon–nucleon interaction) were moving these sub-shells relative to each other. Lately it is considered that nuclear deformation is responsible for these discrepancies.

Above $A \sim 100$ the ground states of odd nuclei do not appear to reflect the order in which the shells should fill; e.g. four isotopes of Sn ($Z = 50$) have ground states $\frac{1}{2}^+$ instead of only one as predicted. The reason lies in the pairing term; adding a second neutron into one of the isotopes $\frac{1}{2}^+$ should produce the pair $(3s_{\frac{1}{2}})^2$, but the pairing of neutrons in a higher lying state with higher l, probably the $1h_{\frac{11}{2}}$-state, could give an increased binding which more than offsets the higher energy of the intrinsic state. In which case the pair of neutrons will be found in the $1h_{\frac{11}{2}}$-state in the next even–even isotope, with the $3s_{\frac{1}{2}}$-state unoccupied. So the next odd isotope of Sn will again be $\frac{1}{2}^+$, etc.

Also, just below this region, ground states $\frac{7}{2}^+$ should be found corresponding to neutron odd numbers 57–63. None is found, but in each nucleus where expected, a low-lying $\frac{7}{2}^+$ state occurs. A similar explanation can probably be made, though it is possible that the ordering of the $1g_{\frac{7}{2}}$- and $3s_{\frac{1}{2}}$-states should be inverted. For the more massive nuclei, nuclear deformation is expected to occur away from the closed shells, and this would produce a different ordering of the ground-state spins (see next section).

A triumph of the shell model concerns the location of 'islands of isomerism'. It had been noticed that long-lived isomers (excited states of nuclei which decay by γ-emission or its equivalent, internal conversion, with lifetimes long enough to be classified as radioactive) tended to cluster at certain regions of the periodic table. To understand this it is necessary to know something about γ-ray selection rules (see Chapter 5). At this stage we merely need to know that the bigger the spin change the slower the transition. In certain regions of the periodic table, the first excited state is expected to differ in spin from the ground state of an odd-A nucleus by a large amount. For example, at $Z = 39$ the last proton is in a $2p_{\frac{1}{2}}$-state, whereas the first proton excitation is $1g_{\frac{9}{2}}$; it lies at 0.381 MeV in ^{87}Y, at 0.91 MeV in ^{89}Y, and at 0.551 MeV in ^{91}Y, giving lifetimes of 14 hours, 16 s, and 50 min respectively. Thus it is seen that the spin-orbit term which produces the right sequence of shells also puts sub-shells of high spin in juxtaposition with sub-shells of low spin and so is responsible for isomerism.

Some mention must be made of the magnetic moments of the odd-A nuclei. From the assumption of maximum pairing, the properties of the nucleus should be those of the odd nucleon. The magnetic moment contributed by a single nucleon is

$$\mu = \mu_{\mathrm{N}}[g_l\langle \boldsymbol{l}\cdot\boldsymbol{j}\rangle + g_s\langle\boldsymbol{s}\cdot\boldsymbol{j}\rangle]\frac{m_{\max}}{j(j+1)} \text{ with } m_{\max} = j,$$

where $2\langle \boldsymbol{l}\cdot\boldsymbol{j}\rangle = j(j+1) + l(l+1) - s(s+1)$, and correspondingly for $\langle \boldsymbol{s}\cdot\boldsymbol{j}\rangle$. This can readily be derived using the vector model (see e.g. Kuhn 1961 and Appendix E). By and large the agreement with experimental data is poor. It is best when the odd nucleon immediately precedes or follows shell closure; examples for the odd proton are $^3_1$H, $^{15}_7$N, $^{39,41}_{19}$K, $^{89}_{39}$Y, and for the odd neutron $^3_2$He, $^{17}_8$O, $^{207}_{82}$Pb, but even here $^{41}_{20}$Ca, $^{45}_{21}$Sc, $^{91}_{40}$Zr show large deviations. $^{19}_9$F might have been quoted above were it not for the fact that theoretical computation has come up with an untidy, three-particle mixture of $s_{\frac{1}{2}}$, $d_{\frac{5}{2}}$, and $d_{\frac{3}{2}}$ wavefunctions outside an ^{16}O core, that happens to give the magnetic moment of an $s_{\frac{1}{2}}$ proton. The deviations of magnetic moments are as yet unexplained; they could arise from three sources: (1) strict pairing does not occur, (2) nuclei may not be spherical, and (3) the pion field around a nucleon inside a nucleus should differ from that around a nucleon in isolation; this could give rise to a magnetic moment differing from that obtained by vector addition of the individual moments as ascribed to isolated nucleons.

Finally, how does the shell model cope with excited states? The general ordering of levels indicated in Fig. 3.4 should represent filled levels up to a certain value and single particle excitations beyond this. If strict pairing does occur, the excited states of an odd-A nucleus should follow this sequence of excitation. Such single-particle excitations do occur, especially when they refer to large l-values in regions of sub-shells of low l. However, the number of excited states up to, say, 3 MeV above ground is usually far greater than the number of single-particle levels available. We are forced to conclude that strict pairing has broken down (it may well take only ~1 MeV to break a pair) and that extra states are obtained by the coupling of three or more particles. An extreme example of this is ^{19}F. Its first excited state, at 108 keV, is $\frac{1}{2}^-$ whilst the lowest single-particle states of odd parity, $(1f_{\frac{7}{2}})$ and $(2p_{\frac{3}{2}})$, should be several MeV higher. Merely decoupling the last two neutrons will not produce odd parity since all three particles are in even-parity orbits—the $s_{\frac{1}{2}}$, $d_{\frac{5}{2}}$, and $d_{\frac{3}{2}}$ which lie close together. It has therefore been assumed that the closed proton p-shell of the ^{16}O core has been disrupted, and one of the protons has gone into the s-d shell to create something like an α-particle; the $(p_{\frac{1}{2}})$ 'hole' provides the odd parity for this state and others at somewhat higher excitation. Thus, where excited states are concerned, even the closed shells must be considered vulnerable.

The shell model forms a working basis for attempting to determine the level scheme of a nucleus, but there are interactions between the nucleons, termed 'residual forces' since they remain after accounting for the well by averaging out the main body of the forces. These residual forces are often arrived at empirically in building up nuclei progressively from a double-closed shell core by comparison with experimentally determined level schemes.

Collective motion

In this and the previous chapters, reference has been made to a collective nuclear motion. For this to occur, motion of closed shells is implied, and this in turn means that such shells cannot be spherically symmetric—quantum mechanically there can be no motion about an axis if the system has symmetry about that axis. Thus the occurrence of collective motion is characterized by a departure from spherical symmetry of the nuclear system; it may be the rotational motion of a permanently deformed nucleus or the motion of a nucleus vibrating (possibly about an undeformed mean structure) through different degrees of deformation.

The possibility of permanent deformation was investigated by Nilsson, who calculated the single-particle states in a spheroidal potential well having a symmetry axis. The distortion of the spherical well is then characterized by the single parameter $\delta = \Delta R/R$, as defined in Chapter 1. For such a well *fixed in space* it must be appreciated that, because of the absence of spherical symmetry, the total angular momentum J is no longer a constant of the motion, but because the system does have symmetry about one of its principal axes the component of angular momentum along this axis, designated by K, is indeed a constant of the motion. The behaviour with δ of the low single-particle levels is shown in Fig. 3.5, where the value of $|K|$ is shown on each curve. Since the deformation is quadrupole no distinction is made between up and down, so that each curve has degeneracy 2 corresponding to $\pm K$. In the limit $\delta = 0$, the spherical states appear; in particular the $d_{\frac{5}{2}}$-state appears as the confluence of three states of different $|K|$ and therefore has degeneracy 6, which is in accord with its spatial degeneracy, $2j + 1$.

If the problem is made physically more realistic by isolating the well in space rather than fixing it, then the well itself will be capable of rotation, but the rotation must have no component along the symmetry axis. If the rotational motion is slow compared to the single-particle motion (the adiabatic limit), then the latter will have a well-defined angular momentum component along the nuclear axis K, but j itself will not be well defined. The total angular momentum J will also be well defined (an isolated body) and will have a component K along the nuclear axis and a component M along the external axis of quantization. J will be a vector sum of j and R, the nuclear rotation (see Fig. 3.6). The rotational energy will be given by

$$E_{\mathrm{R}} = (\hbar^2/2I)\langle R^2 \rangle = (\hbar^2/2I)\langle (J - j)^2 \rangle$$
$$= (\hbar^2/2I)(J(J + 1) + \langle j^2 \rangle - 2\langle J \cdot j \rangle)$$
$$= (\hbar^2/2I)(J(J + 1) + \langle j^2 \rangle - 2K^2), \qquad (3.1)$$

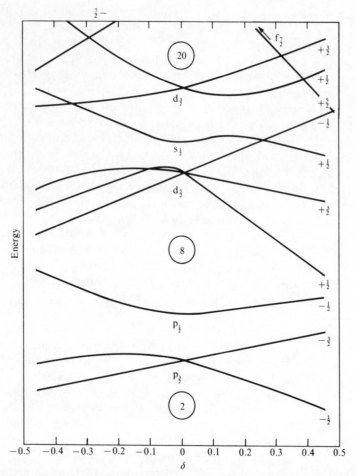

Fig. 3.5.   Single-particle states in a fixed ellipsoidal potential well, characterized by $\delta = \Delta R / R$. The states are labelled with $|K|$ and parity (they are doubly degenerate) when $\delta \neq 0$. At $\delta = 0$ they correspond to the sub-states of the usual shell-model states. The $1S_{\frac{1}{2}}$-state has been omitted. (Based on Nilsson (1955). *Dan. mat. – fys.medd.* **29,** No. 16.)

since it can be shown that

$$\langle J_{x'}j_{x'} + J_{y'}j_{y'} \rangle = 0$$

unless there are components of $K \pm 1$ as well as K in the wavefunction. Since there are components of $\pm K$ due to up-down degeneracy, these terms are not zero for $|K| = \frac{1}{2}$, but

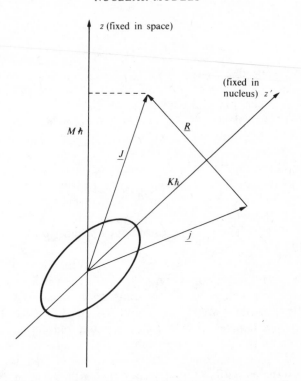

Fig. 3.6. Coupling scheme for a deformed nucleus. $\boldsymbol{J} = \boldsymbol{j} + \boldsymbol{R}$, $j_{z'} = J_{z'} = K\hbar$, $J_z = M\hbar$. Note that $\boldsymbol{j}$ itself is not a constant of motion, nor is j^2 well defined; e.g. $J^\pi = \frac{7}{2}^-$ can be constructed from $f_{\frac{7}{2}}$, $h_{\frac{9}{2}}$, $h_{\frac{11}{2}}$ orbitals coupled to $R = 0$ or 2, and, except for extremes of coupling, all components will be present.

this case will be ignored. The term $\langle j^2 \rangle$ can be evaluated in a more detailed treatment; obviously it must be greater than or equal to K^2 since the component of j along the axis of symmetry has the fixed magnitude K. However for a given wave function relative to the nuclear axis, it is possible to build up a sequence of states of different J starting from the lowest value, $J = K$ (J cannot be smaller than one of its components), and increasing in unit steps. These states have different rotational energies, as given by the expression above, but the same energy due to basic structure. The term $\langle j^2 \rangle$ is therefore unchanged in going from one value of J to another and so does not affect the spacing of levels within the rotational band; the levels in the band can be located relative to the lowest member by the expression

$$E_J = (\hbar^2/2I)[J(J + 1) - K(K + 1)].$$

Thus we started with a state in the fixed well which was doubly degenerate ($\pm K$) and from it, on releasing the well and isolating the system in space, built up a rotational band the lowest member of which has $J = K$ and therefore has degeneracy $(2K + 1)$. Thus the $d_{\frac{5}{2}}$ configuration, which originally split into three doubly degenerate states in the fixed well has now produced three states of $J = \frac{1}{2}$, $\frac{3}{2}$ and $\frac{5}{2}$ in the isolated system. In this case the confluence at $\delta = 0$ creates difficulty since it can be only six-fold degenerate, whereas the converging J states have the total degeneracy of 12—and states cannot be created or destroyed by changing a perturbation, but must only be moved around.

The reason for the discrepancy is that the rotation of the system (and the rotational energy) has not been included in Fig. 3.5 since the well is fixed, must be included in the more realistic situation in which the system is isolated in space. How to evaluate this depends upon the nature of the rotation. It could be a rigid-body rotation, or an effective rotation of a hydrodynamical fluid, an example of which is the tidal deformation of the earth by the moon. On the surface of the earth this appears like a rotation, but the motion is in fact up and down. Experimental investigation indicates that the latter is nearer the truth and $I \sim I_0\delta^2$, where I_0 is the rigid-body moment of inertia of a sphere, namely, $\frac{2}{5}MR^2$. The rotational energy therefore becomes very large at small δ so the states $J = \frac{1}{2}$ and $\frac{3}{2}$ produced from a $d_{\frac{5}{2}}$ particle move up to very high energies as δ becomes small. More correctly, the coupling scheme breaks down; it is no longer valid to couple the single-particle motions rigidly to the well and then to allow the whole system to rotate. This adiabatic model only approximates to reality when δ is sufficiently large for the rotational energy to be small compared to the spacing of the different particle configurations.

At sufficiently large δ, the diagram correctly predicts the ordering of states. The particles couple strongly in pairs with equal and opposite K, so an even–even nucleus will always have $K = 0$ and a ground state $J = 0$. Putting $J = 0$, $K = 0$ in the rotational energy gives a rotational band of levels of energies, relative to the ground state, $J(J+1)\hbar^2/2I$, where because of symmetry requirements (up $\equiv$ down) only even values of J can occur. This symmetry requirement is met by expressing the wavefunction of the system as a combination of the wavefunctions $(J, M, +K)$ and $(J, M, -K)$ thus: $(\psi^J_{M,+K} + (-1)^{J-K}\psi^J_{M,-K})$. For $K = 0$ it is seen that the two components are the same except for the term $(-1)^J$ which eliminates all odd J. For other values of K no such restriction arises, so J increases in unit steps up a band. For odd-A nuclei, the sequence of levels in Fig. 3.5 determines a value of $|K|$ for the ground state. A rotational band is now built up having energies relative to ground $\hbar^2\{J(J+1) - K(K+1)\}/2I$. It should be noted that $K = \frac{1}{2}$ produces a more complicated ordering of states which will not be discussed, except to mention that it arises from the $\langle \mathbf{J} \cdot \mathbf{j} \rangle$ term in the rotational energy.

A typical level scheme is shown for an even and an odd nucleus in Fig. 3.7. It will be seen that rotational states up to large values of J are obtained and that they tend to be lower

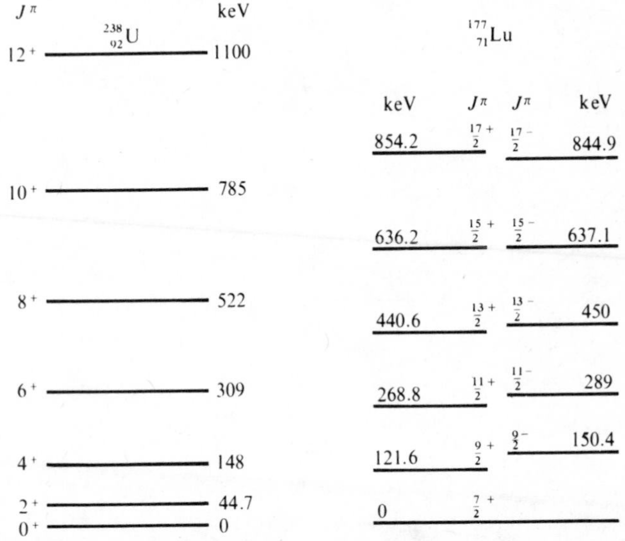

Fig. 3.7. Examples of rotational bands in even-A and odd-A nuclei. In the latter nucleus the level schemes representing the two non-interacting bands have been separated laterally. For clarity some levels unconnected with the bands have been omitted from both schemes.

lying than the next particle configuration. Their locations usually conform well with the simple expressions above, but there is usually a small discrepancy which increases with J and which is ascribed to an increase of I due to a centrifugal stretching of the nucleus. Experimentally, levels in a rotational band are revealed by the very large E2 transitions (see Chapter 5) between members—the intrinsic strength can be as large as 200 times single-particle strength in heavy, deformed nuclei in the rare-earth region. Such large enhancements arise from the rotation of the nucleus as a whole, corresponding to a correlated motion of up to Z protons and therefore capable in principle of producing enhancements up to Z^2.

Finally, nuclei undeformed in the ground state can have excited states corresponding to vibrations through varying

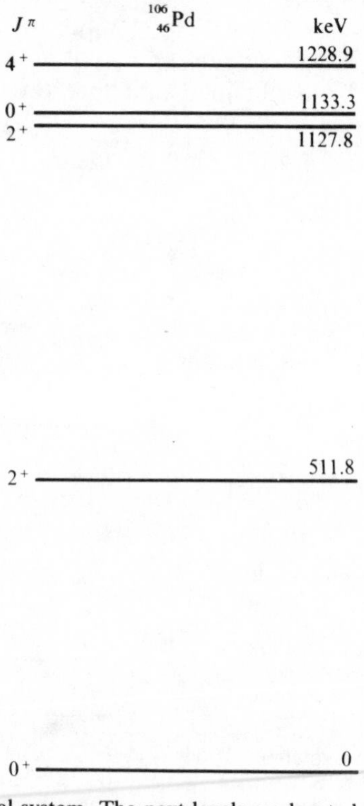

$J\pi$ $^{106}_{46}$Pd   keV

4^+ ———————————— 1228.9

0^+ ———————————— 1133.3

2^+ ———————————— 1127.8

2^+ ———————————— 511.8

0^+ ———————————— 0

Fig. 3.8.   A vibrational system. The next levels are located around 1.6 MeV and could include members of the three-quantum vibrational group of states.

degrees of deformation. As in the case of the harmonic oscillator, the different vibrational states are equally spaced, but because the deformation is quadrupole, the vibrational quanta are characterized by 2^+ giving for an even–even nucleus the sequence of states (0^+), (2^+), $(0^+, 2^+, 4^+)$, $(0^+, 2^+, 3^+, 4^+, 6^+)$, etc. where the brackets correspond to 0, 1, 2, 3, etc. vibrational quanta coupled using Bose statistics. The bracketed states are degenerate in the model, but departure from the model removes the degeneracy as Fig. 3.8 shows. Because the couplings of vibrational quanta produce the gamuts of states extending from low J values upward, the onset of states produced by multiple excitations of shell model states quickly disturbs the sequence by mixing states of like J^π. This is quite different from the behaviour of rotational states which are still pure states at much higher excitation energies—see p. 133.

The optical model

The previous models refer to a static or quasi-static system since the excited states so far introduced are bound states decaying only by the relatively weak radiation process and therefore having lifetimes much longer than a characteristic nuclear time, which is the time taken for a nucleon to traverse a nuclear diameter. In this situation the quantum-mechanical problem consists of setting up stationary states (just as for the ground state) and then treating the radiative process as a perturbation. When the energy of the system is large enough to permit nucleon (or composite particle) emission then the decay is via the strong interaction so the state can have a lifetime of the order of the characteristic time. The decay process is no longer a small perturbation and a different approach is needed. The optical model meets this need.

This model attempts to cope with the quantum-mechanical problem of the scattering and absorption of particles impinging on a nucleus. Since a simple potential has been used to represent the nucleus in the shell model, it is appealing to do the same for the scattering problem. An argument for the use of such a potential was based upon the long mean free path of a nucleon inside the nucleus as a consequence of the Pauli principle. However, a nucleon coming in from outside has $(E_{\text{kin}} + 7)\,\text{MeV}$

(the incoming kinetic energy + binding energy) more energy than any nucleon within the nucleus. It can collide with such a nucleon thereby losing energy and exciting that nucleon. Because of the Pauli principle it cannot interact with nucleons lying more than $(E_{kin} + 7)$ MeV below the Fermi surface of the nucleus. Thus as the kinetic energy of the incident nucleon increases, the limitations of the Pauli principle are reduced and the mean free path of this nucleon gets shorter.

How is this concept fed into the problem? A beam of particles traversing a region in which the mean free path of a particle is λ will decrease in beam intensity with distance as $I = I_0 \exp(-x/\lambda)$. If the beam of particles, without absorption, is represented by the plane wave amplitude $\sqrt{I_0} \exp(ikx)$, then with absorption it should be

$$\sqrt{I_0} \exp(-x/2\lambda)\exp(ikx) = \sqrt{I_0} \exp\{i(k + i/2\lambda)x\}.$$

The wavenumber has therefore been modified from k to $(k + i/2\lambda)$. But $\hbar k = \{2m(E - V)\}^{\frac{1}{2}}$, where E is the total energy of the particle and V the potential energy. Putting in the modified expression gives

$$k^2 - (1/2\lambda)^2 + ik/\lambda = 2m(E - V')/\hbar^2.$$

Since the total energy of the particle must be real (it can move outside the region of absorption to be measured), an imaginary component must be added to the potential. Expressing this as $V + iW$ gives $2m(E - V)/\hbar^2 = k^2 - (1/2\lambda)^2 \approx k^2$ for reasonably large λ; and $2mW/\hbar^2 = -k/\lambda$ connects the imaginary component with the mean free path.

In its early stages therefore the optical model used a complex square-well potential, $-(V + iW)$ for $r < R$, zero for $r > R$. This was later changed to the Saxon–Woods potential

$$\frac{-(V + iW)}{1 + \exp\{(r - R)/c\}},$$

where V, W, R, and c were looked upon as arbitrary parameters in fitting the experimental scattering data at a given energy. The latter two parameters should obviously correspond in some way to the density parameters; the simplest way being to assume that the potential follows the density when allowance is made for the range of nuclear forces. The real term V might be expected to approach the shell-model potential at low enough energy. There is no good reason why V and W should behave in the same way; indeed, an energy-dependence of W is expected from the previous argument which is unlikely to be similar to that of V. Also it can be argued that the bound nucleons near the nuclear surface should be those with greatest energy, for which collisions with the incoming nucleon are least restricted. Thus a peaking of W near the surface may be more realistic, and this can be achieved by associating W with the derivative of the Fermi (Saxon–Woods) function, or by simply using a Gaussian function peaked at the surface. In addition, for treatment of polarization data, it has been found necessary to introduce a spin-orbit coupling term, $\propto \boldsymbol{l} \cdot \boldsymbol{s}$, where $\boldsymbol{l}$ and $\boldsymbol{s}$ refer to the incident nucleon.

It is instructive to consider S-wave scattering from a square-well potential of radius R in the limit of very small W. In order to avoid the complexities of determining elastic scattering cross-sections, the variation with energy of the reaction cross-section is estimated; for $W = 0$, there is no absorption and therefore only elastic scattering, but for W small enough the simplifying features of $W = 0$ can be used. The wave function inside the well (for $W = 0$) is of the form $(A/r)\sin Kr$, satisfying the requirement of a node at the origin in r times the radial wavefunction, whilst outside the well the solution can be written $(B/r)\sin(kr + \delta)$, where K and k are the interior and exterior wavenumbers: $K = 2m(E + V)^{\frac{1}{2}}/\hbar$ and $k = (2mE)^{\frac{1}{2}}/\hbar$. The standing-wave solutions indicate that no absorption has taken place, i.e. as many neutrons are moving away from the nucleus as toward it (and the flux of neutrons toward the nucleus is proportional to $k|B|^2$, though the direct connection is complicated since the neutrons proceeding toward the nucleus constitute a plane wave).

The continuity of the wave function and its derivative at the nuclear surface lead to the equations

$$k \tan KR = K \tan(kR + \delta)$$

and

$$A = Bk/(K^2\cos^2 KR + k^2 \sin^2 KR)^{\frac{1}{2}}.$$

If for simplicity it is assumed that $K > k$ then A/B is seen to vary between maxima of ~ 1 when $KR \sim (n + \frac{1}{2})\pi$ and minima of

$\sim k/K$ when $KR \sim n\pi$. Also $|A|^2$ will fall to half its maximum value when $KR \sim (n + \frac{1}{2})\pi \pm k/K$.

If now a very small imaginary term is introduced into the potential, then absorption will take place within the nucleus, but for small enough W this will hardly change the wave functions. From the connection between W and mean free path it is obvious that the reaction yield will be proportional to $W|A|^2$. The reaction cross-section will therefore be proportional to $W|A|^2/k|B|^2 = Wk/(K^2\cos^2 KR + k^2\sin^2 KR)$, showing maxima and minima with variation of neutron energy. If, as in Chapter 4, the problem were approached in a different way, namely, by inserting the neutron inside the nucleus and observing its transition out, then it will be seen that the maxima correspond to the formation of 'virtual states' of the system. The reaction is said to exhibit 'resonance' when the neutrons are at an energy appropriate to one of the virtual states within the nucleus, cf. forced oscillations of a pendulum near its natural frequency or of an electric circuit. The ratio of resonance half-width to resonance spacing can be estimated from the above equations as $(2k/K) \times 1/\pi$ or $(4k/K) \times 1/2\pi$. The ratio has been expressed in this latter way since the factor $4k/K$ is the mismatch factor at the potential step (see Chapter 4 and Problem 4.1), whilst the factor $1/2\pi$ represents an important theorem concerning widths of virtual states, namely, that on average the decay width of a state into any one channel, after allowing for matching the internal and external wave functions, should be $1/2\pi$ times the local spacing of levels with the same spin and parity. For a medium-sized nucleus and a well-depth of 50 MeV, the level spacing of s-wave single-particle neutron states ($J^\pi = \frac{1}{2}^+$ for $J^\pi = 0^+$ targets) is about 15 MeV and the single-particle width about 2 MeV. Thus the single-particle resonances are broad but well separated.

In going to a more realistic model of the nucleus, as well as the single-particle states there are also found vast numbers of levels of the same spins and parities but corresponding to excitations of more than one particle. In fact the single-particle states alone cannot produce reaction but only re-emission of the same particle. That reaction does take place indicates that the single-particle state is mixed in (by some unspecified perturbation) with all these more complicated states and W must in some way represent the degree of mixing. In the mixed-up system the theorem concerning level width and spacing will on average still hold (see below), though the mixing may reduce the spacing from a few MeV to a few eV.

Inclusion of a large imaginary term in the potential will broaden out the resonance even more (the absorption term is representing other channels into which the state can decay, so the width should roughly increase by the amount W, as seen in Chapter 6). Since absorption has taken place, followed by decay into another (reaction) channel, a broad resonance is to be expected in this reaction channel, e.g. the emission of inelastic neutrons. Data on total neutron cross-sections are presented in Fig. 3.9 and also the results of analysis in terms of the optical model. A feature of the model is that the parameters which describe the potential should vary only slowly from nucleus to nucleus, and this appears to be the case. Typical values for the potential for neutron scattering from a wide range of nuclei at

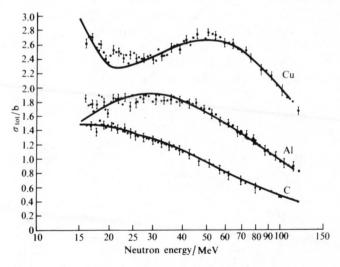

Fig. 3.9. Plot of total neutron cross-section against energy for C, Al and Cu. The full lines show optical-model fits. (Based on Bowen *et al.* (1961). *Nucl. Phys.* **22**, 640.)

neutron energies up to ~50 MeV are

$$V = 52.5 - 0.3E_n \text{ (MeV)},$$
$$W = 2.5 + 0.3E_n \text{ (MeV)},$$
$$R = r_0 A^{\frac{1}{3}}, \text{ where } r_0 = 1.20 - 1.25 \text{ fm},$$
$$c = 0.5 - 0.6 \text{ fm}.$$

In addition the spin–orbit interaction referred to previously in this section takes on the value: $V_{so} = \left(\dfrac{\hbar^2}{m_\pi c}\right)^2 U_s \left(\dfrac{1}{r}\dfrac{df}{dr}\right) \boldsymbol{l} \cdot \boldsymbol{s}$, where $f(r)$, with r expressed in fm units, represents the Saxon–Woods shape with the parameters as above, and U_s takes the value 11 MeV. The factor $\left(\dfrac{\hbar}{m_\pi c}\right)^2 = 2(\text{fm})^2$ serves to keep the dimensions correct. Like some representations of W, V_{so} is made to peak at the nuclear surface.

The success of the model is illustrated better by measurements of angular distributions as shown in Fig. 3.10. The situation is paralleled in optics by diffraction through a circular aperture or, better still, the scattering (diffractive) of a small transparent drop of liquid of refractive index different from unity. Also, perhaps paradoxically, the model has had a great deal of success in the analysis of slow neutron data, where compound-nucleus (Chapter 6) resonances of width typically measured in fractions of 1 eV are

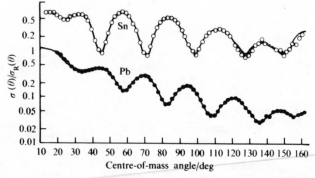

Fig. 3.10. Differential cross-section (expressed as a ratio to Rutherford) for elastic scattering of 30.3 MeV protons, showing optical-model fits. (Based on Satchler (1967). *Nucl. Phys.* **A92**, 273.)

observed. Representing the nucleon–nucleus interaction by a simple central potential cannot possibly account for these manifestations of the detailed structure of the nucleus, but it has been shown that the optical model predicts average cross-sections where the range of averaging covers many compound nucleus resonances. For W very large (~single-particle spacing) the ratio of level width (without barrier) to spacing, which determines the average cross-section, will be of the order of $1/2\pi$; but for W small this ratio, known as the 'strength function', will follow the shape of the optical model resonance. See also Appendix G.

Problems

3.1. The lowest states in $^{207}_{82}$Pb are (energy in MeV in brackets) $\frac{1}{2}^-$ (0), $\frac{5}{2}^-$ (0.57), $\frac{3}{2}^-$ (0.90), $\frac{13}{2}^+$ (1.63), and $\frac{7}{2}^-$ (2.34). Interpret these in terms of the shell model.

3.2*. $^9_4$Be, $^{13}_6$C, $^{17}_8$O have ground state magnetic moments -1.17 nm, $+0.70$ nm, and -1.89 nm respectively. Deduce values for (a) the spins and parities, (b) the magnetic moments of these nuclei, as predicted by the shell model. Comment on any discrepancies.

3.3*. Under what circumstances are (a) the single-particle shell model, (b) the rotational model expected to give a good description of the properties of nuclei. Describe two features of nuclear behaviour for which either model makes characteristic predictions and show how these predictions can be tested experimentally.

3.4*. The energy levels of the $K = 0$ ground-state rotational band of ^{238}U are given in Fig. 3.7 (p. 63). Estimate the moment of inertia I by equating the excitation energies with rotational energies. Why does I increase with increasing angular momentum? Why does the sequence contain only even values of J? Compare I with $I_{\text{rigid}} = \frac{2}{5}AMR^2$ for the appropriate rigid-sphere rotation and comment on the difference.

3.5. Use Bose statistics to show that three 2^+ vibrational quanta can couple to form the states $J^\pi = 0^+$, 2^+, 3^+, 4^+, 6^+.

3.6. The optical model potential for neutrons on a medium-weight nucleus ($A \sim 125$) is given by the combinations $(0, 50, 2)$, $(30, 40, 8)$, $(100, 25, 8)$, and $(300, 10, 5)$, where

the first number in a bracket is the neutron energy, the
second $-V_0$, and the third $-W_0$ all in MeV). Determine
the mean free path at each energy near the centre of the
nucleus; show that the nucleus is reasonably transparent at
each end of the range, and even at maximum absorption
there is ~10 per cent chance of a neutron passing
diametrically through the nucleus. Explain the general
trend of V and the mean free path. If the nucleus is
transparent for near-thermal neutrons, how is it that they
can be captured for long periods of time (see Chapter 6)?

3.7. Work through the problem outlined on p. 67 and verify the
two equations given there.

4. Spontaneous decay of nuclei I: α and β decay

In Chapter 1 it has been shown that the stable nuclei are surrounded by neighbours which are unstable to β-decay, whilst further away from the stability line nucleon emission can take place. These latter nuclei can only be observed in nuclear reactions. For mass values greater than 150, even the nuclei on the stability line become unstable, not to nucleon emission but to emission of α-particles because the binding energy per nucleon of the α-particle is similar to that of a nucleus.

α-Decay

Historically, α-particle decay was observed in the heavier nuclei ($A > 206$), and the lifetime for decay was seen to be very strongly correlated with the energy of the α-particle emitted (because the process is a break-up into two particles, the energy of the α-particle is well defined for unique parent and daughter states). For these nuclei all observed α-emissions lie within the energy limits $4 \, \text{MeV} < E_\alpha < 9 \, \text{MeV}$, the lower value being associated with a lifetime $\sim 10^{+18} \, \text{s}$ and the upper value with $10^{-6} \, \text{s}$, according to the Gieger–Nuttall rule which can be expressed as $\ln \lambda = a + b \ln E$, where λ^{-1} is the mean life. This rule explains why, although almost all nuclei in the range $A = 150$–200 are unstable to α-emission, only a few have been observed to decay in this way; in general the available energy is very low and the lifetime too long to permit detection of the process. An example of an observed decay in this region is $^{152}_{64}\text{Gd}$, with a lifetime of 10^{14} years and an α-particle energy $2.24 \, \text{MeV}$. These figures are not compatible with the Geiger–Nuttall relationship as applied to the natural radioactive series, but it will be seen later that the connection between lifetime and energy depends very strongly on the nuclear charge Z.

Another historic experiment was the bombardment of the daughter nucleus with α-particles of the same energy as (or even greater than) those emitted by the parent. The scattering was observed to be of the pure Rutherford type, which, on the face of it, indicated that the nucleus was still behaving effectively as a point in its interaction with an α-particle approaching it from outside, even though this particle was closer to it than the region from whence (in the classical sense) an α-particle must have originated in the decay of the parent. This paradox was explained by Gamow in his theory of barrier penetration.

In Fig. 4.1 an idealized form is presented of the potential to which an incoming or outgoing α-particle is exposed. The shape of the potential is dominated by the Coulomb interaction at long distances and by nuclear forces at short distance. These latter have been simplified to a square-well form, though they should conform to something like the optical potential, which will smooth out the sharp discontinuity and also, by means of the imaginary term, produce absorption within the nucleus. Also shown in Fig. 4.1 is a cruder approximation to the potential in terms of potential steps.

It is instructive to solve the wave-mechanical problem, in one dimension, of a plane wave of wavenumber K impinging from the left onto this latter potential, being partially transmitted, and

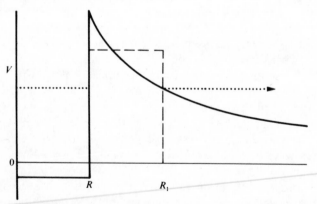

Fig. 4.1. Idealized potential for s-wave α-particle plus product nucleus. Also shown is the potential step approximation.

finally emerging as a wavenumber k in the far right region; in the intermediate region it must be represented by real exponentials (multiplied by complex amplitudes) if, in this region, its total energy is less than the height of the potential step. This exercise is left to the reader.

If in this problem the incident wave is of the form $I \exp(iKr)$ and the transmitted wave $A \exp(ikr)$ then the transmission coefficient T is defined by the ratio of transmitted to incident flux

$$T = v_O |A|^2 / v_I |I|^2 = k |A|^2 / K |I|^2,$$

where v_I, v_O are ingoing and outgoing velocities. For this particular problem

$$T = \frac{16Kk\gamma^2}{(K^2 + \gamma^2)(k^2 + \gamma^2)} \exp(-2\gamma\Delta), \qquad (4.1)$$

where $\gamma = \{2m(V - E)\}^{\frac{1}{2}}/\hbar$ and Δ is the width of the barrier, $(R_1 - R)$.

There is a direct correspondence between the solution of a one-dimensional problem and the radial dependence of the solution of the corresponding three-dimensional problem with spherical symmetry. The one-dimensional wave function $U(r)$ is simply connected to the radial dependence of the three-dimensional wave function $F(r)S(\theta, \phi)$ by $U(r) = rF(r)$, but the potential for the one-dimensional problem is not just $V(r)$ but $V(r) + l(l + 1)\hbar^2/2mr^2$, i.e. allowance must be made for the centrifugal potential. Thus the solution above is appropriate for transmission of an $l = 0$ α-particle through the spherically symmetric potential step. Also the transmission coefficient is unchanged since the flux at r is $v(r)r^2 |F(r)|^2$ in the three-dimensional case and therefore proportional to $v |U|^2$ as before. It has been assumed that $S(\theta, \phi)$ is correctly normalized, i.e. $\int |S(\theta, \phi)|^2 d\Omega = 1$.

However, considered as a three-dimensional problem, there must be a node (in $U(r)$) at the origin. In principle, no simple solution as obtained can give such a node since, to conserve flux, the reflected wave on the left (i.e. inside the spherical potential) must have smaller intensity than the incident wave. But the difference ($\sim\exp(-2\gamma\Delta)$) may be very slight, and, to a good approximation, can be ignored. The requirement of a node at the origin in addition to the (amplitude and) phase relationship between the two components derived from solution of the previous problem overdetermines the solution within the potential, leading to a discrete set of solutions corresponding to energy levels of the α-particle. These solutions are not quite stationary states since they are decaying via transmission through the barrier.† However, if $\exp(-2\gamma\Delta)$ is small then the decay is slow when measured on a nuclear time scale. The decay constant λ_α (in s^{-1}) will be the outgoing flux when the wave function inside the 'nucleus' of radius R is normalized to unity:

$$\int_0^R |I \exp(iKr) + J \exp(-iKr)|^2 \, dr = 1,$$

where $J \exp(-iKr)$ is the reflected wave. Because of the node at the origin the integral becomes $\int_0^R 4 |I|^2 \sin^2 Kr \, dr$ giving $|I|^2 \approx 1/2R$, since the point R is very close to being an antinode. From the uncertainty principle, or, more strictly, from a Fourier Transform of a decaying state to its representation as a spread of frequencies and therefore energies,

$$\Gamma_\alpha = \hbar\lambda_\alpha = \hbar v_0 |A|^2 \quad \text{when} \quad |I|^2 = 1/2R,$$

that is

$$\Gamma_\alpha = \frac{\hbar^2 k}{m} \frac{|A|^2 K |I|^2}{K |I|^2} = \frac{\hbar^2}{m} T \frac{K}{2R} = \frac{\hbar^2}{2mR^2} \quad (KRT). \qquad (4.2)$$

This expression separates the α-decay width into two parts: the part $\hbar^2/2mR^2$, which has dimensions of energy (width) and is the nuclear scale factor for width, and the dimensionless bracketed term, which is the transmission coefficient suitably modified for an effectively bound α-particle rather than a flux. This transmission factor is the product of a major term, the barrier penetrability, and a minor term which here represents the

---

† A more accurate solution of the problem introduces difficulties of normalization. The outgoing wave must slowly increase in amplitude as the distance from the nucleus increases, representing the fact that, at an earlier time, the probability of finding an α-particle within the nucleus was greater, and therefore the leakage through the barrier was greater. Thus the normalization integral diverges outside the nucleus. The approximate solution given can be looked upon as a perturbation approach in which an infinite confining potential has been switched off at $t = 0$ and the rate of increase of the final state (corresponding to a free α-particle) is determined when the initial state is known to represent one α-particle confined within the nucleus.

product of two mismatch factors arising from the two discontinuities in the potential (cf. reflection of optical waves at a glass surface or vibrations at a joint between stretched strings). In a more realistic calculation, introducing the Coulomb potential, it is reasonable to expect (and more detailed calculations confirm this) the penetrability term to be replaced by $\exp\{-2\int_R^{R_1} \gamma(r)\,dr\}$, since γ is now a varying function of r. The limits of integration, R and R_1, are the points where $\gamma(r) = 0$. The slowly varying γ effectively changes adiabatically to an equally slowly varying k without discontinuity and so eliminates the mismatch factor at R_1. At R the discontinuity of the nuclear square-well potential will introduce a mismatch factor, but a (rapidly changing) Saxon–Woods-type potential will reduce the mismatch though not necessarily eliminate it. This accounts for the fact that analysis in terms of a square well appears to require a potential of approximately zero inside the nucleus whereas analysis in terms of the optical model usually gives a negative internal potential ($-25\,\text{MeV}$ or so), though with considerable latitude since there are other variables in the analysis. Note that the modified barrier penetrability may also be applied to the case $l \neq 0$, since $\gamma(r)$ will be computed with the centrifugal barrier included.

The chief source of uncertainty in the calculation of decay widths and lifetimes is in deciding what fraction of the characteristic width $\hbar^2/2mR^2$ must be used for real nuclei. The original calculations employed the full value, as given here, which for nuclei with $A > 200$ is about $0.2\,\text{MeV}$, though the precise value will depend on the nuclear potential adopted. This value implies that one alpha particle effectively exists in, and is trying to get out of, the nucleus at all times. It was later suggested that $1\,\text{eV}$ may be more realistic; this estimate was based upon widths of resonances in slow neutron reactions (see Chapter 7). Such a large reduction in nuclear factor was offset in the barrier-penetrability term by employing a larger nuclear radius. Nowadays, these radii so deduced are considered to be too large, which is to be expected since the levels from which the width was inferred are at an excitation of order $7\,\text{MeV}$ and statistical theories of nuclear levels require that as level densities increase then level widths (for a particular channel, though not necessarily total widths) should diminish. Somewhere near one

TABLE 4.1

l	0	1	2	3	4	5	6	7	8
Reduction factor	1	0.7	0.37	0.14	0.04	7×10^{-3}	1×10^{-3}	1×10^{-4}	7×10^{-6}

per cent of the characteristic width is now considered a reasonable estimate. Obviously, in the face of such uncertainties, variations in mismatch factors due to details of the nuclear potential are of little significance.

An examination of the α-spectrum of a given radioactive nucleus reveals in general more than one transition energy. The ground state of the parent nucleus can decay to a number of states in the daughter nucleus (of which the ground state may be one). In general too the higher the excitation of the daughter state the smaller the intensity of the transition, since the penetrability is so strongly dependent upon the α-energy. For the naturally occurring α-emitters a rough estimate is a factor of 3 decrease in transition probability for 100 keV decrease in α-energy. In addition, the angular momentum requirements will contribute to intensity variations. Table 4.1 gives a rough guide to this effect.

As an example consider the decay $^{238}_{94}\text{Pu} \rightarrow {}^{234}_{92}\text{U} + \alpha$. Table 4.2 gives the measured transitions to states of known J (since $^{238}_{94}\text{Pu}$ has $J = 0$ in the ground state, the final J value is also the l-value for the emitted α-particle).

The first five transitions are to members of a rotational band. Such states have similar wave functions since they represent the same nuclear configuration in different degrees of rotation; it is therefore gratifying to see that the decrease in transition strength is largely accounted for by the variation of penetrability with energy and angular momentum. The sixth transition (others are given in the literature for this decay, but this one is typical) has a much weaker strength than would be accounted for by these

TABLE 4.2

J	0^+	2^+	4^+	6^+	8^+	1^-
E (keV)	0	43	143	297	499	786
Relative strength:						
Estimated	1	0.25	9×10^{-3}	3×10^{-5}	3×10^{-8}	1×10^{-3}
Observed	1	0.39	1.2×10^{-3}	7×10^{-5}	1×10^{-8}	2×10^{-7}

means. It therefore reflects the nuclear contribution to the process. Thus the conclusion to be drawn is that for members of a rotational band (or for that matter a vibrational band) the simple estimates of relative strengths should be fairly good, but in general the dominant factor is the nuclear intrinsic width when low-lying states are being compared.

Finally, mention must be made of 'long-range' α-particles (this historical term is synonymous with 'high-energy'). An excited state can, in principle, emit α-particles; indeed its decay probability into the α-channel will probably be greater than that of the ground state. However, it will be competing unfavourably with γ-emission (see Chapter 5) at least for states near ground; at high enough excitation the reverse will occur but such states are not excited in the natural α-decay series. If the energy available to the α-particle is very large then the decay probability will be correspondingly large. If also the γ-transition probability is reduced because of high spin change, for example, then it may be possible that the excited state can have an observable α-decay. The key word is observable; modern techniques have revealed quite large numbers of 'long-range' transitions giving low-intensity α-groups above an intense ground-state group. From what has been said they are most likely to be found in conjunction with short-lifetime ground-state decays. An example is $^{212}_{84}$Po, with a ground-state half-life of 304 ns and $E_\alpha = 8.79$ MeV, having three such transitions, of which the most intense, 1.8×10^{-4} of the ground state strength, is from a state at an excitation of 1.8 MeV.

β-Decay

In Chapter 1, β-decay has been introduced as a process involving the simultaneous emission of a β-particle and a neutrino. The neutrino was postulated to satisfy the requirements of conservation of energy, momentum and angular momentum in this process. The β-spectrum is continuous up to an upper limit which is the energy expected for the β-particle in the absence of an undetected particle. Similarly experiments involving the direction of recoil of the daughter nucleus indicate that momentum is not conserved—or rather that an undetected particle has also been emitted. Also, to conserve angular momentum a neutron cannot decay into two spin-$\frac{1}{2}$ particles, whether isolated or inside a nucleus. The neutrino must be neutral since charges balance without it. It must have intrinsic spin $\frac{1}{2}$ and a very small rest mass, since the β-energy at maximum accounts very closely for all the energy available. Efforts were made in the past to detect the neutrino, but failed. Now that recent efforts have been successful, it is not surprising that former methods failed since it is now known that neutrinos can pass through the Sun, the centre of which is a prolific source—indeed neutrino emission accounts for a considerable fraction of the energy emitted by stars.

A simple theory of β-decay was provided by Fermi, who exploited the similarity with emission of radiation. A proton (or neutron) changes into a neutron (or proton) by simultaneously creating the $\beta - \nu$ pair just as an excited state of an atom changes to a lower state by creating a photon. Note that in isolation a proton cannot decay to a neutron because it has less mass, but inside a nucleus a bound proton may be in a state of higher energy than an equivalent bound neutron, so the process can take place; if the total masses are favourable then the reaction can go.

Considering first the decay of an isolated neutron ($n \rightarrow p + e^- + \bar{\nu}_e$) then the wavefunctions of the two light particles can be represented by plane waves in the directions of emission, $\psi_\beta = N_\beta \exp(i\mathbf{k}_\beta \cdot \mathbf{r}_\beta)$ and $\psi_\nu = N_\nu \exp(i\mathbf{k}_\nu \cdot \mathbf{r}_\nu)$. The origin of the coordinate system is assumed to be the location of the nucleon both before and after the event (i.e. recoil has been neglected for simplicity). The plane-wave approximation should be adequate for the neutrino, but for the β-particle the long-range Coulomb interaction with the proton (or nucleus when at a later stage the nucleons are considered to be part of a nucleus) will distort the wave. This point will be taken up later. In the absence of guidance, Fermi assumed the simplest possible interaction namely that it is proportional to the simultaneous overlap of all four particles, i.e.

$$H_{\mathrm{fi}} = g \int \psi_{\mathrm{f}}^* \psi_{\mathrm{i}} \, \mathrm{d}\tau = g \psi_\beta^*(0) \psi_\nu^*(0)$$

in this simple case of point nucleons at the origin. From perturbation theory (Fermi's golden rule) the transition probability per unit time is

$$P = \frac{2\pi}{\hbar} |H_{fi}|^2 \frac{dn}{dE},$$

where dn/dE is the density of final states. In applying this formula it is convenient to enumerate the final states when the system is confined to a specified (large) volume Ω. The normalization factors for the plane waves, N_β and N_ν, are then both $1/\Omega^{\frac{1}{2}}$, and the density of plane-wave states of a particle having momentum between p and $(p + dp)$, with the particle anywhere in Ω is $p^2 \, dp \, \Omega/2\pi^2\hbar^3$. The number of states for the β-momentum in the range p_β to $(p_\beta + dp_\beta)$ and the ν-momentum in p_ν to $(p_\nu + dp_\nu)$ is then

$$dn = (p_\beta^2 \, dp_\beta/2\pi^2\hbar^3) \cdot (p_\nu^2 \, dp_\nu/2\pi^2\hbar^3)\Omega^2.$$

In multiplying these two functions together it has been assumed that there is no correlation between the directions of emission of electron and neutrino. This is partly because of the initial simple assumptions, but chiefly because the final nucleon is available to take up the requisite momentum required for conservation; because it is so massive it can do this and still have little effect on the energy balance. It does not appear directly in the calculations, but its presence has been felt nevertheless.

To determine dn/dE it is necessary to transform from variables p_β, p_ν to p_β, E, using the equation

$$E = c(p_\nu^2 + m_\nu^2 c^2)^{\frac{1}{2}} + E_\beta,$$

where E is the total available energy. The transformation is

$$dp_\beta \, dp_\nu = \frac{\partial p_\nu}{\partial E} dp_\beta \, dE = \frac{1}{v_\nu} dp_\beta \, dE = \frac{1}{c} dp_\beta \, dE,$$

since the rest mass of the neutrino is either zero or close to it. Very close to cut-off of the β-spectrum this neglect of the rest mass of the neutrino could distort the spectrum, but such effects have not yet been established with complete certainty. Thus the density of states is given by

$$\frac{dn}{dE} = \frac{\Omega^2 p_\beta^2 p_\nu^2}{4\pi^2\hbar^6 c} dp_\beta = \frac{\Omega^2 p_\beta^2 (E - E_\beta)^2}{4\pi^4\hbar^6 c^3} dp_\beta,$$

using $E_\nu = cp_\nu = E - E_\beta$.

The transition probability per unit time to produce electrons in the momentum range p_β to $(p_\beta + dp_\beta)$ is

$$P(p_\beta)\,dp_\beta = \left(\frac{g^2}{2\pi^3\hbar^7 c^3}\right)(E - E_\beta)^2 p_\beta^2\,dp_\beta, \qquad (4.3)$$

or in terms of energy

$$P(E_\beta)\,dE_\beta = P(p_\beta)\frac{dp_\beta}{dE_\beta}\,dE_\beta = \left(\frac{g^2}{2\pi^3\hbar^7 c^3}\right)(E - E_\beta)^2 \frac{p_\beta^2}{v_\beta}\,dE_\beta. \qquad (4.4)$$

Note that the result is independent of Ω, the volume in which the system was confined, since $|H_{\text{fi}}|^2 = g^2/\Omega^2$; it is essential that this should be the case since the confining volume has been introduced merely to simplify the enumeration of the density of final states. In the non-relativistic approximation the energy spectrum is proportional to $(E - E_\beta)^2\sqrt{E_\beta}$, and this accounts for the general shape of a β-spectrum, namely an abrupt rise from $E_\beta = 0$, with a vertical tangent at the origin, to a maximum at $E_\beta = E/5$, falling smoothly to zero at $E_\beta = E$ and meeting the axis with a horizontal tangent.

To extend this treatment to complex nuclei it is necessary to consider the decay of each nucleon as an independent event and to sum the amplitudes of each event. So far the interaction is at a point, so the $\beta-\nu$ combination cannot take away orbital angular momentum; since no mention has been made of spin the combination must be assumed to be the spin $= 0$ state. In the nucleus the decaying nucleon merely changes its charge but remains in the same nuclear state—the instantaneous effect can be represented by $\sum_k \tau_k^\pm \psi_i$, where $\tau_k^\pm$ is the isospin raising or lowering operator acting on the kth nucleon in the nuclear wave function ψ_i (τ^- for β^- emission, τ^+ for β^+). This instantaneous wave function must be projected onto the final wave function to determine the nuclear contribution to the process, giving the matrix element

$$M_{\text{fi}} = g \int \psi_f^* \sum_k \tau_k^\pm \psi_i\,d\tau.$$

The summation may be taken over all nucleons since τ^- acting on a proton gives zero—it changes neutrons to protons—and correspondingly for τ^+. Obviously, from what has been said above, the selection rule $\Delta J = 0$, with no parity change, holds for the nuclear states.

Extending the theory to include the possibility of the $\beta-\nu$ combination taking away one unit of spin leads to a parallel decay obeying the selection rule $\Delta J = 0$, ± 1, with no parity change (but $J = 0 \rightarrow 0$ excluded). The nuclear matrix element must contain a spin operator capable of changing the spin orientation of the nucleon k—it must be in the form of a vector in order to take away (spin) angular momentum. There is no need to go into its form here; the simple picture of the $\beta-\nu$ combination being emitted in a wave having $L = 0$, $S = 1$ is sufficient to arrive at the above selection rule. The spinless emission is known as the Fermi process and the spin-1 emission is ascribed to Gamow–Teller. The processes so far discussed are termed 'allowed'.

The treatment of nucleons as point particles is an approximation which allowed the $\beta-\nu$ combination to take away no orbital angular momentum. Whilst there is little to be gained by giving the nucleon structure in the decay of an isolated neutron (to preserve parity a two-unit change would be necessary and this is unlikely for such a small structure), it should be noted that even for point nucleons it may be possible to change the orbital angular momentum of a nucleus. If the emitting nucleon is at the periphery of the nucleus then, classically, its recoil momentum will produce a change of angular momentum. The greatest change will occur when an electron of maximum energy is emitted tangentially from the surface, and will be of order $(E_\beta/c)R_N$. Taking $E_\beta \sim 1\,\text{MeV}$ (for which $E_\beta/c = p_\beta$ is only approximate) and $R_N \sim 6\,\text{fm}$ (a medium-size nucleus) gives $0.03\hbar$ for the classical angular momentum in this extreme situation. Thus the taking away of one unit of angular momentum must be looked upon as possible but improbable. Mathematically, the two plane waves representing β and ν appear in the interaction integral as an effective plane wave $\exp(\mathrm{i}\boldsymbol{k} \cdot \boldsymbol{r})$ where $\boldsymbol{k} = \boldsymbol{k}_\beta + \boldsymbol{k}_\nu$ and $\boldsymbol{r} = \boldsymbol{r}_\beta = \boldsymbol{r}_\nu$, the common point of interaction. This effective plane wave is expanded within the confines of the nucleus giving $\exp(\mathrm{i}\boldsymbol{k} \cdot \boldsymbol{r}) = 1 + \mathrm{i}\boldsymbol{k} \cdot \boldsymbol{r} - \frac{1}{2}(\boldsymbol{k} \cdot \boldsymbol{r})^2 \ldots$. The matrix element for the process will now become

$$H_{\mathrm{fi}} = g \int \psi_{\mathrm{f}}^* \sum_k \tau_k^\pm (1 + \mathrm{i}\boldsymbol{k} \cdot \boldsymbol{r}_k \ldots) \psi_{\mathrm{i}}\, \mathrm{d}\tau,$$

where for convenience the possible spin factor is omitted. If ψ_f and ψ_i have opposite parities then the leading term in the bracket will integrate to zero, but the next term, which is itself odd, may not do so. If polar axes are chosen such that $\boldsymbol{k}$ lies along z then $\boldsymbol{k} \cdot \boldsymbol{r}_k$ becomes $kr_k \cos \theta$. The angular wave functions of nucleon k in ψ_f and ψ_i are in the form of spherical harmonics $Y_{l,m}(\theta, \phi)$, and their properties are such that $\int Y_{l_f,m_f}^*(\theta, \phi) \cos \theta \, Y_{l,m_i}(\theta, \phi) \, d\Omega$ will be zero unless $m_f = m_i$† and $l_f = l_i \pm 1$. Thus the kth nucleon changes angular momentum by one unit (cf. emission of a photon by an electron in an atom) and, by recoupling to the rest of the nucleus, the selection rule follows: $\Delta J = 0, \pm 1$, with change of parity ($J = 0 \to 0$ excluded) for the Fermi process. By coupling up to $S = 1$, the corresponding Gamow–Teller selection rule is $\Delta J = 0, \pm 1, \pm 2$, with change of parity.

Consideration of the radial integral indicates that these transitions will be down by a factor $(kR_N)^2$ on the allowed transition—the k^2 term arises from the integral and the R_N^2 from the fact that an integral $\int \psi_f^* r \psi_i r^2 \, dr$ must be $\sim R_N \int |\psi_f| |\psi_i| \, r^2 \, dr$, and we assume that all integrals of the latter type have approximately the same value. These types of transition are known as 'first forbidden' since they are down by the factor $(kR_N)^2$, which, as previously estimated, is $\sim 10^{-3}$ or 10^{-4}. The next term in the series would produce second forbidden transitions and will be down by the same factor again.

Before examining whether this classification conforms with experimental fact, it is necessary to consider the effect of nuclear charge on the process since this will vary from nucleus to nucleus. The wave function of a particle in a Coulomb field is complicated, but here we merely need to know the behaviour at $r = 0$, where the process occurs, and $r = \infty$, where the products are observed. If at infinity the wave function for the β is the original plane wave then $|\psi_\beta(0)|^2$ will be multiplied by the factor $F(Z, E) = 2\pi\eta \{1 - \exp(-2\pi\eta)\}^{-1}$, where $\eta = \pm Ze^2/4\pi\varepsilon_0 \hbar v_\beta$ (+ for electrons, − for positrons), v_β being the speed at infinity and Z the atomic number of the final nucleus. The Coulomb correction enhances the probability of electron emission and decreases that of positron emission, especially at low energies. In

† A different choice of axes results in $m_f = m_i \pm 1$.

the latter case it is effectively the Gamow barrier-penetrability factor since the positron may well be created with negative kinetic energy and need to penetrate the barrier before emerging with positive energy. The enhancement of the β^--process occurs because the attractive potential increases $|\psi_\beta(0)|^2$ relative to $|\psi_\beta(\infty)|^2$. The normalization of the wave is largely determined by $|\psi_\beta(\infty)|^2$ since Ω is very much larger than the nuclear volume, but the transition probability depends on $|\psi_\beta(0)|^2$. These effects are illustrated in Fig. 4.2(a) and Coulomb distorted spectra in Fig. 4.2(b).

With the Coulomb correction the momentum spectrum for allowed transitions is of the form

$$P(p_\beta)\,\mathrm{d}p_\beta = CF(Z, E_\beta)(E - E_\beta)^2 p_\beta^2\,\mathrm{d}p_\beta, \qquad (4.5)$$

where the constant C contains terms independent of the β-energy. Thus a plot of $\{P(p_\beta)/Fp_\beta^2\}^{\frac{1}{2}}$ against E_β is a straight line intersecting the abscissa at E. This, or its equivalent for the energy spectrum, is known as a Kurie plot, an example of which is shown in Fig. 4.3. The linear plot results from allowed trans- itions only; in forbidden transitions terms in $k^2 = k_\beta^2 + k_\nu^2 + 2k_\beta \cdot k_\nu$ appear, which depend upon the relative direction of the $\beta-\nu$. On integrating over all directions, the term $k_\beta \cdot k_\nu$ averages to zero, and the two remaining terms inject a factor $\{p_\beta^2 + (E - E_\beta)^2/c^2\}$ into the momentum spectrum for a first forbidden transition. The Kurie plot will not be linear, but a modified Kurie plot, including this factor, can restore the straight line. This tech- nique can be used to determine the type of decay (see Fig. 4.3).

A convenient classification of β-transitions is based on the lifetime of the decay

$$\frac{1}{\tau} = \int P(p_\beta)\,\mathrm{d}p_\beta = g^2\,|M_{\mathrm{fi}}|^2 f,$$

which defines f as an integral over all the energy dependent terms in the spectrum, and contains all the fundamental constants excluding the coupling constant. Thus for an allowed transition

$$f_0 = (2\pi^3 c^3 \hbar^7)^{-1} \times \int_0^{p_{\max}} (E - E_\beta)^2 p_\beta^2 F(Z, E_\beta)\,\mathrm{d}p_\beta.$$

The product $f_0 t_{\frac{1}{2}}$, where $t_{\frac{1}{2}}$ is the half-life ($= \tau \ln 2$), is therefore a

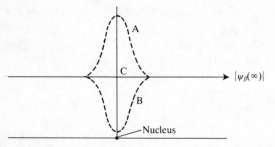

Fig. 4.2(a)   A represents $|\psi|$ for a β^- distorted plane wave; the nuclear attraction enhances $|\psi|$ in its vicinity.
B represents $|\psi|$ for a β^+ distorted plane wave; the nuclear repulsion diminishes $|\psi|$ in its vicinity.
C represents an undistorted plane wave.
Since the restraining volume Ω can be chosen such that $R_\Omega \gg R_N$, the normalization factor, $\Omega^{-\frac{1}{2}}$ for the undistorted wave, is unaffected by the local changes of ψ.

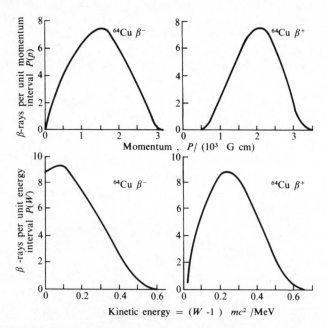

Fig. 4.2(b)   Effect of Coulomb potential on β-spectra for ^{64}Cu decay, which shows β^- and β^+ (and electron-capture) decay. Note that the β^- energy spectrum has finite intensity at zero kinetic energy, but the momentum spectrum does not (the effect of the relationship $dE = v\,dp$). (Based on Evans (1955). *The atomic nucleus*, McGraw-Hill, New York.)

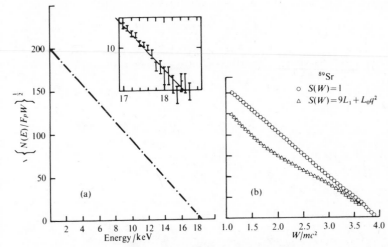

Fig. 4.3. (a) Kurie plot for allowed decay of tritium. Inset shows fit near end point. Lack of deviation indicates that the rest mass of the neutrino must be small. (Based on Lewis (1970). *Nucl. Phys.* **A151**, 120.) (b) Kurie plot for first forbidden decay of ^{89}Sr. The curved plot is the usual plot with f_0 in the ordinate function; the linear plot contains a theoretical correction for the first forbidden transition. (Based on Wohn and Talbert (1970). *Nucl. Phys.* **A146**, 33.)

measure of $(g^2 |M_{fi}|^2)^{-1}$ for an allowed transition and is found to vary between 10^3 and 10^6 in magnitude. Transitions having values near 10^3 are known as 'super-allowed' since they correspond to (almost) complete overlap of the nuclear wave functions. They are usually Fermi transitions between states of the same isospin multiplet (e.g. β-decay between mirror nuclei), but there exist a few Gamow–Teller transitions in very light nuclei where the overlap is also complete. The remainder of the range corresponds to a diminished overlap. The *same* product $f_0 t$ for a first forbidden transition will be of the form $\{g^2(|M_{fi}|^2/R_N^2)(\overline{kR_N})^2\}^{-1}$, where the bar denotes a suitable average over the β-spectrum. For complete overlap this is up by a factor $(kR_N)^{-2}$ in value on an allowed transition giving a range of order 10^6–10^{10}. Similarly, a second forbidden transition will give $f_0 t$ values $>10^{10}$. Thus there is no strict division between the different classes, but for certain ranges of $f_0 t$ values it is possible to deduce the type with a reasonable degree of confidence, see Table 4.3.

Note that these last considerations should really apply only to β^- transitions since, as has been mentioned in Chapter 1, an

TABLE 4.3

Distribution of $\log_{10} f_0 t$ values for Allowed and Forbidden $\beta^{\pm}$-transitions. The sub-division of forbidden transitions into nonunique and unique categories is referred to later in the chapter—see p. 91. (Taken from Gleit *et al.* Beta decay transition probabilities. *Nucl. Data. Tab. Supp.* Nov. 1963).

$\log f_0 t$	Super-allowed†	Allowed	First forbidden		Second forbidden	
			Nonunique	Unique	Nonunique	Unique
2.8–3.2	6					
3.3–3.7	25‡	1				
3.8–4.2	2	8				
4.3–4.7		57				
4.8–5.2		120	8			
5.3–5.7		99‡	5			
5.8–6.2		74	24			
6.3–6.7		47	57			
6.8–7.2		25	40	1		
7.3–7.7		20	57‡	1		
7.8–8.2		9	38	28		
8.3–8.7		5	25	35‡		
8.8–9.2		9	21	10		
9.3–9.7			3	7	1	
9.8–10.2			5	2		
10.3–10.7		1	2	1	1	1
10.8–11.2			1	2	3	1
11.3–11.7			1		2	1‡
11.8–12.2			5		4‡	1
12.3–12.7			2		3	2
12.8–13.2					3	
≥13.3			1		3	
Total	33	475	295	87	20	6

† We include here also the $0^+ \leftrightarrow 1^+$ transitions in He^6, C^{10}, O^{14}, F^{18}, and Ne^{18}.
‡ Mean value.

alternative decay mode to β^+ is available. This will be discusssed briefly in the next section.

Electron capture

Since atomic nuclei are in general surrounded by atomic electrons an alternative process to β^+-decay can take place, namely,

$$p + \beta^- \rightarrow n + \nu_e \quad \text{(within the nucleus)}.$$

This process is of greatest importance at low available energies—indeed there is a range of available energies (see Chapter 1) up

to 1.02 MeV over which β^+-decay is impossible and only
e^--capture can take place. Also it is expected to be largest for
K-electrons (provided the energy balance is not critical), since
they have strongest overlap with the nucleus, and for nuclei of
large Z, since the electron orbits are more tightly bound, again
leading to greater overlap.

Applying Fermi's golden rule to this process, the density-of-
states function will be that of the neutrino alone—the nucleus is
constrained to recoil in the opposite direction and both particles
have fixed energy. The probability of finding the electron at the
origin of the coordinate system is now no longer inversely
proportional to the volume of the enclosure but to the volume of
the K-orbit (for K-capture), so once again Ω cancels (it has also
disappeared from the density-of-states function for the electron
which is now unity—a defined state). If E is used, as before, to
denote the maximum available energy for the β^+-decay, then the
energy available to the neutrino will be $(E + 1.02 - B_K)$ MeV,
where B_K is the binding energy of the K-electron. There is now
no integral to perform since no variation is allowed of the
products, and

$$\frac{1}{\tau_K} = \frac{2\pi}{\hbar} g^2 |M_{fi}|^2 \frac{1}{\pi} \left(\frac{Z}{a}\right)^3 \frac{(E + 1.02 - B_K)^2}{2\pi^2 \hbar^3 c^3}, \qquad (4.6)$$

where a is the (hydrogen) Bohr radius, so that a/Z is the mean
radius of the K-electron wave function in the parent nucleus, and
where the matrix element M_{fi} is the same as before, and $1/\tau_K$ is
the probability per second of K-capture. Thus the probability of
the process increases as Z^3, whilst that for β^+-decay decreases
with Z (due to the Coulomb effect); for light nuclei β^+-decay will
dominate except for an energy range in E from the negative
value $(-1.02 + B_K)$ MeV to slightly above zero. But for high Z,
K-capture is overwhelmingly predominant for β^+-energies commonly encountered.

Relativistic theory and parity violation

The account of β-decay so far described has been essentially
non-relativistic (NR) in its approach, apart from the use of a
relativistic equation connecting momentum and energy—a trivial
aspect. For relativistic invariance it is essential that time and

space be treated on an equal footing, and this is not the case with the Schrödinger wave equation, which is of second order in spatial derivatives and first order in the time derivative. Putting $E = i\hbar\, \partial/\partial t$, $p_x = (\hbar/i)\, \partial/\partial x$ into the relativistic (R) equation $E^2 = c^2 p^2 + m_0^2 c^4$ gives the Klein–Gordon equation which is second order in all four derivatives. However, its solutions are found to conform to Bose statistics, e.g. photons, and the equation cannot be used for spin-$\frac{1}{2}$ particles. Dirac looked into the conditions under which this second-order equation can be factorized to two first-order equations and further examined the solutions of the resulting equation. He found that the solutions were four-component vectors in a space which was closely allied to the two component space of a NR spin-$\frac{1}{2}$ particle. Indeed two of the components are down by a factor v/c on the other two, such that in the limit $v \to 0$, the R system reduces to the NR system. In addition the 'small' components have opposite parity to the 'large' components, creating the possibility of parity violation at relativistic energies.

In β-decay, the approximation $v/c \ll 1$ is rarely valid for the electron and never so for the neutrino—even for tritium decay the β-particle has maximum kinetic energy of 18.6 keV corresponding to $v/c \sim 0.27$. Thus the product $\psi_\beta \psi_\nu$, which in the NR limit gives four terms which can be regrouped into a scalar ($S = 0$) and a three-vector ($S = 1$) gives in the R limit sixteen terms which can be regrouped into two scalars, two four-vectors, and a six-component antisymmetric tensor. The two vectors are respectively polar and axial; the former changes sign under the parity operation whilst the latter does not, e.g. r and p are polar three-vectors whilst $r \wedge p$ is an axial three-vector. Of the two scalars one is a pseudo-scalar in that it changes sign under the parity operation; an example of such a scalar is the triple three-vector product $a \cdot (b \wedge c)$, where a, b, c are polar three-vectors. Such a product can be used classically (e.g. to represent a volume) but the sign property is usually ignored.

Thus a full relativistic treatment is much more complicated than the picture presented here and has, built into it, terms which mix the parity of the relativistic components. It is possible to choose an interaction in such a way as to eliminate the parity-violating component, and this was believed to be the case until specific measurements were made to test parity conservation in β-decay. Such measurements are designed to determine whether pseudo-scalars are necessary in the description of the reaction. Such a scalar product is $\boldsymbol{\sigma} \cdot \boldsymbol{p}$, where $\boldsymbol{\sigma}$ represents the NR spin of the electron and $\boldsymbol{p}$ its momentum, or $\boldsymbol{J} \cdot \boldsymbol{p}$, where $\boldsymbol{J}$ is the spin of the initial nucleus and the β-momentum $\boldsymbol{p}$ is now measured relative to the same axis as $\boldsymbol{J}$. Obviously, if parity is conserved then $\langle \boldsymbol{\sigma} \cdot \boldsymbol{p} \rangle$ and $\langle \boldsymbol{J} \cdot \boldsymbol{p} \rangle$ must be zero. The first measurements were made on the decay of ^{60}Co, with the initial state polarized by cooling in a magnetic field. It was then found that the number of β-particles directed into the hemisphere with a component along the field was different from that directed into the opposite hemisphere, indicating that $\langle \boldsymbol{J} \cdot \boldsymbol{p} \rangle$ was not zero. The final conclusion to be drawn from such measurements was surprising, not so much because parity was not conserved but because its violation appeared to be the maximum possible when extrapolated to the limit $v \rightarrow c$.

When polarization measurements are ignored as in the simple determination of the spectrum, then the simple predictions of the Fermi theory are largely correct, but, in more detail, separation into 'large' and 'small' terms of the nucleon wave functions is necessary. Part of a first forbidden transition can arise from the 'small' term of an allowed transition, and vice versa, though in the latter case the effect will not be discernible since it is also reduced by the factor $(kR_N)^2$ relative to the main component; in the former case the factor serves to enhance this extra component. Thus, whilst under most circumstances the simple theory copes with allowed transitions, great caution is needed in extending its results to forbidden transitions. It is now convenient to subdivide such transitions further into nonunique and unique classes. For a first forbidden process having $\Delta J = 0$ or ± 1 (and change of parity) the accompanying 'small' process is an allowed transition with its corresponding enhancement, whereas if $\Delta J = \pm 2$ the 'small' process will be second forbidden and therefore quite negligible. The latter transitions are termed unique and are expected to conform with the simple NR theory of first forbidden transitions; the former transitions are termed nonunique and are expected to fill in the gap between $f_0 t$-values for allowed and first forbidden transitions of the NR theory according to the size of the 'small' terms in the Dirac wave functions. Table 4.3 confirms this interpretation of the data.

Problems

4.1. Solve the one-dimensional problem of transmission through the rectangular potential as given in Fig. 4.1 (p. 74). Show that in the limit of the barrier width approaching zero, the transmission becomes that calculated for a potential step.

4.2. Solve the integral $\int_R^{R_1} \gamma \, dr$ (see p. 77) for the more realistic potential of Fig. 4.1., where the limits of integration are the classical distance of closest approach and the nuclear radius (use the transformation $r = r_0 \cos^2\theta$). Show that, in the limit $E_\alpha \ll$ barrier height, the Gamow factor $\exp(-2\pi Zze^2/4\pi\varepsilon_0\hbar v)$ is the important term in S-wave barrier penetration.

4.3. Show that the non-relativistic formula for the β-spectrum for an allowed transition gives a mean kinetic energy of the β of $\frac{1}{3}$ the maximum energy, whilst the relativistic formula, for a very high maximum energy, gives a factor $\frac{1}{2}$. Derive expressions for the maximum recoil energy of the nucleus of mass A when the β is emitted at the mean energy in both cases. (Neglect Coulomb effects.)

4.4*. Derive an expression for the electron momentum spectrum in allowed β-decay (ignore Coulomb corrections). Find the approximate dependence of the total decay rate on the maximum electron momentum p_0 when $p_0 \gg mc$.

   Compute the value of the partial decay rate for the pion decay $\pi^+ \rightarrow \pi^0 + e^+ + \nu_e + 4.5$ MeV given that the Fermi decay of ^{14}O to ^{14}N (excited) has a momentum end point $p_0 = 2.26$ MeV/c and a mean lifetime of 103 s.

4.5*. Explain the meanings of the following terms: (a) first forbidden, allowed and super-allowed β-transitions, and (b) Fermi and Gamow–Teller matrix elements. Classify with regard to (a) and (b) the following: (i) $n \rightarrow p$; (ii) $^6_2\text{He}(0^+) \rightarrow {}^6_3\text{Li}(1^+)$; (iii) $^{14}_8\text{O}(0^+) \rightarrow {}^{14}_7\text{N}(0^+)$, $f_0 t = 3.3 \times 10^3$; (iv) $^{35}_{16}\text{S}(\frac{3}{2}^+) \rightarrow {}^{35}_{17}\text{Cl}(\frac{3}{2}^+)$, $f_0 t = 1.0 \times 10^5$; (v) $^{36}_{17}\text{Cl}(2^-) \rightarrow {}^{36}_{18}\text{Ar}(0^+)$.

4.6. ^7_4Be decays by electron capture chiefly to the ground state of ^7_3Li, and the recoil energy of the ^7Li has been measured to be 57 eV to 1 per cent accuracy. If the mass difference between the two atoms is 0.862 MeV/c^2 show that the result is consistent with a neutrino rest mass anywhere in the range of zero to $\sim 10\,\text{keV}/c^2$.

4.7. It is stated on p. 80 that the neutrino has spin $\frac{1}{2}$. Determine the selection rules for β-decay if the neutrino had spin $\frac{3}{2}$. Hence show that the spin cannot be $\frac{3}{2}$ (or greater).

5. Spontaneous decay of nuclei II: electromagnetic transitions

Thus far the discussion has been restricted to the decay of nuclear ground states, apart from the occasional intrusion of an excited state decaying, for example, as an isomeric state, or giving a 'long-range' α-particle. In the next chapter states at high enough excitation to emit nucleons will be considered, but here the electromagnetic decays of states within 1 MeV or so above the ground state are treated. The electromagnetic process is, in general, a stronger one than the weak β-process, so that such excited states only rarely β-decay, though there are a few instances when the excited states lie close to the ground.

The treatment of radiation presented here is largely based upon the corresponding process of electric dipole radiation from excited atomic states. Later in the chapter it will be shown that such radiation can take place only between states satisfying the criteria $\Delta J = 0, \pm 1$ and change of parity. But, experimentally, using techniques to be described briefly in the next chapter, it is found that in nuclei γ-transitions frequently take place between states not conforming to these criteria. It is therefore necessary to extend the treatment to cover the case of emission via the higher electric multipoles and via magnetic multipoles. Such transitions can be ignored in atoms since they occur so seldom and then only under very special conditions. If the rapid process of electric dipole (E1) radiation cannot take place, then the state usually de-excites by collision. In nuclei, on the other hand, it transpires that E1 emission is not a great deal faster than the other processes, and in any case most low-lying states have the same parity as each other since they probably arise from different couplings of the same shell model particles or from collective motion. But nuclei are isolated from each other by virtue of their

94

charge, and are coupled only weakly to the atomic electrons; a state once formed can only decay via the electromagnetic process, though it need not necessarily lead to γ-emission since the Coulomb coupling to the atomic electrons can result in the emission of an electron termed, misleadingly, a 'conversion electron'. This latter process is electromagnetic and gives the same information concerning nuclear states as does γ-emission.

γ-Radiation

The difficulty in calculating the spontaneous decay of an excited state is that one does not know the perturbing term in the wave equation which is responsible for the transition. However, the reverse process of absorption can be treated by using the electric vector of the radiation as a time-dependent perturbation. Einstein, using a statistical argument (known as 'Einstein's As and Bs,), obtained a relationship between the two processes. A fuller account is given in Pauling and Wilson (1935), but very briefly the argument is as follows.

The coefficient of absorption is defined such that the probability of a transition from a lower state n to a higher state m is given by $B_{n \to m} \rho(v_{mn})$, where $\rho(v_{mn}) \, dv$ is the energy density of radiation in the range v_{mn} to $v_{mn} + dv$ with v_{mn} the frequency of the transition, i.e. $v_{mn} = (E_m - E_n)/h$. The probability of transition down is given by $A_{m \to n} + B_{m \to n} \rho(v_{mn})$, where $A_{m \to n}$ is the coefficient of spontaneous emission and $B_{m \to n}$ is the coefficient of induced emission. If we consider the equilibrium of this two-state system in a bath of radiation at infinite temperature then $B_{n \to m} = B_{m \to n}$, since the two states will be present in equal proportions, and the induced processes will be much more probable than the spontaneous process. It is assumed that the two states are non-degenerate; this can be achieved by applying a magnetic field to separate out sub-states of n and m—the transitions between a single sub-state in each level is then under consideration. If now the equilibrium is considered at a finite temperature, then the ratio of abundances will be given by the Boltzmann factor, $N_m/N_n = \exp\{-(E_m - E_n)/kT\}$, whilst $\rho(v)$ is given by Planck's radiation law,

$$\rho(v) = (8\pi h v^3/c^3)\{\exp(hv/kT) - 1\}^{-1};$$

but at equilibrium

$$N_n/N_m = \{A_{m\rightarrow n} + B_{m\rightarrow n}\rho(\nu_{mn})\}\{B_{n\rightarrow m}\rho(\nu_{mn})\}^{-1}.$$

From these equations

$$A_{m\rightarrow n} = (8\pi h\nu_{mn}^3/c^3)B_{n\rightarrow m}$$
$$= (2\hbar\omega_{mn}^3/\pi c^3)B_{n\rightarrow m}$$

It is not proposed to go through the complete calculation of $B_{n\rightarrow m}$ as given in the above reference for E1 transitions, but to show how to modify this approach in order to produce the other multipoles. The perturbation resulting in absorption was taken to be due to the electric vector of the radiation field, $E_0\exp\{i(\mathbf{k}\cdot\mathbf{r} - \omega t)\}$. Since $k = mv/\hbar = E/\hbar c$ for a photon, at ~1 MeV and a nuclear radius of 5 fm then $kR_N \sim \frac{1}{40}$. Thus the simplification $E_0\exp(-i\omega t)$ is a fairly good one, though not so good as in the atomic case. Now a uniform electric field is given by a dipole at infinity, so this approximation, giving a perturbing potential $-eE_0\cdot\mathbf{r}\exp(-i\omega t)$, leads to E1 absorption of amplitude proportional to the matrix element $\int\psi_m^*(\sum_i ez_i)\psi_n\,d\tau$, where the summation i is taken over the protons of the nucleus and where E_0 has been taken as directed along the z-axis. To go from the matrix element to the transition probability introduces a number of summing or averaging processes. In applying the thermodynamic argument the nuclear system is interacting with isotropic radiation, so integration over all directions of the wave vector $\mathbf{k}$ is required and also averaging over polarizations of the radiation. In addition the matrix element as defined refers to specific initial and final substates, so summation over substates must be made (strictly speaking, summing over final states but averaging over initial states).

To quote the final result

$$A_{m\rightarrow n} \approx \frac{1}{4\pi\varepsilon_0}\left(\frac{4\omega_{mn}^3}{3\hbar c^3}\right)D_{mn}^2, \tag{5.1}$$

where D_{mn} now represents the scalar magnitude of the dipole moment and is independent of the particular substates of the initial and final states; it is the radial part of the (substate dependent) vector dipole moment. For a single particle transition it is

$$D_{mn} = \int_0^\infty \varphi_n(r) \, er \, \varphi_m(r) r^2 \, dr$$

where $\varphi_m(r)$ and $\varphi_n(r)$ are (normalized) radial wave functions in the states m and n. This expression ignores the coupling of the spatial states to the spin states for the single particles and, in a more complicated case, the coupling of single-particle states to each other, nor has any attempt been made to cater for the $(2J + 1)$ degeneracies of the states. It is therefore correct only to an order of magnitude—when the Weisskopf single particle estimate is made later it will be realized that, for such a crude model the neglect of the above refinements has little relevance. If however one has a situation in which most of these refinements cancel, then more care is needed. An example is the comparison of measurements of Coulomb excitation with γ-emission between the same states. As will be seen on p. 131, Coulomb excitation is essentially a process of induced absorption, whereas spontaneous emission is proportional to induced emission. The equation $B_{n\to m} = B_{m\to n}$ holds for the individual substates, but for the states a whole $(2J_n + 1)B_{n\to m} = (2J_m + 1)B_{m\to n}$ (obviously the two equations refer to somewhat different Bs—the latter equation requires the substate independent definition as discussed above). This result also follows from Fermi's golden rule and time reversal invariance; the latter makes the matrix element of a process the same as that for the inverse process and the former assigns a density of final states to the rate, i.e. $B_{m\to n} \propto (2J_n + 1)$ and $B_{n\to m} \propto (2J_m + 1)$.

The higher electric multipole contributions can be obtained by using the expansion $\exp(i\mathbf{k} \cdot \mathbf{r}) = 1 + i\mathbf{k} \cdot \mathbf{r} - \frac{1}{2}(\mathbf{k} \cdot \mathbf{r})^2 \ldots$ in the expression for the radiation field. The leading term has given a component of the E1 matrix element; the next term gives a component of the E2 matrix element $\int \psi_m^* ke(\sum_i x_i z_i)\psi_n \, d\tau$ if the propagation vector $\mathbf{k}$ is taken along the x-axis (it must be perpendicular to $\mathbf{E}_0$, since electromagnetic waves are transverse). And similarly for higher orders.

From the matrix elements the selection rules arise. Since z is an odd function, the product $\psi_m^* \psi_n$ must also be odd to give a non-zero dipole term, i.e. there must be a change of parity. Similarly, xz is even on reflection through the origin, so the E2 transition is characterized by no change of parity. In more detail, if the particle making the transition is in a state of orbital angular momentum $(l_i m_i)$ before and $(l_f m_f)$ after the transition, then the angular dependence of the above component of the E1 matrix element will be $\int Y^*_{l_f m_f}(\theta, \phi) \cos \theta Y_{l_i m_i}(\theta, \phi) \, d\Omega$, and the properties of the spherical harmonics (Pauling and Wilson 1935) are such that this integral is zero unless $m_f = m_i$ and $l_f - l_i = \pm 1$ (there are two other components which lead to $m_f - m_i = \pm 1$). Similarly the above component of the E2 matrix element, $\int Y^*_{l_f m_f}(\theta, \phi) \cos \theta \sin \theta \cos \phi Y_{l_i m_i}(\theta, \phi) \, d\Omega$, will be zero unless $m_f - m_i = \pm 1$ and $l_f - l_i = 0, \pm 2$, with $0 \rightarrow 0$ excluded since it cannot satisfy the former condition. Again, there are other components, for which $m_f - m_i = 0, \pm 2$, the xy component giving $\Delta m = \pm 2$.

For nuclei consisting of a single proton outside closed shells, the selection rules on (J, M) will be the same as those obtained on (l, m) for transitions between states corresponding to excitations of this proton, apart from the addition of a spin vector. In the general case it is necessary to decouple the rest of the nucleus from the particle making the transition and then to recouple after the transition. The mathematics of these coupling operations is beyond the scope of this book, but it is not difficult to see that the selection rules will become

E1: $\Delta J = 0, \pm 1$ ($0 \rightarrow 0$ excluded), with parity change,

E2: $\Delta J = 0, \pm 1, \pm 2$ ($0 \rightarrow 0, \frac{1}{2} \rightarrow \frac{1}{2}$ excluded), with no parity change.

The exclusions arise because the radiation field of the multipole of order l must take away angular moment $\hbar l$, so for $l = 2$ the vector equation $\boldsymbol{J}_i = \boldsymbol{J}_f + \boldsymbol{2}$ must hold and it cannot for $0 \rightarrow 0$ and $\frac{1}{2} \rightarrow \frac{1}{2}$.

So far no mention has been made of magnetic multipole transitions for the very good reason that the interaction has been taken to be that of an electric field. But electromagnetic radiation also possesses a magnetic vector $\boldsymbol{H}_0$, perpendicular to both $\boldsymbol{E}_0$ and $\boldsymbol{k}$. Again the leading term will arise by assuming the (magnetic) field constant over the nucleus giving rise to the perturbation term $-\boldsymbol{\mu} \cdot \boldsymbol{B}_0 \exp(-i\omega t)$, where the magnetic moment $\boldsymbol{\mu} = \mu_N \sum_i (g^i_s \boldsymbol{s}_i + g^i_l \boldsymbol{l}_i)$, where μ_N is the nuclear magneton and g^i_s, g^i_l are the appropriate spin, orbital g-factors for the nucleon i. To simplify matters, consider the case of a single neutron outside a closed-shell core then $\boldsymbol{\mu} \cdot \boldsymbol{B}_0 = \hbar^{-1} g_s \mu_N (s_x B_{0x} + s_y B_{0y} + s_z B_{0z})$, where s_x, s_y, s_z are spin operators (see Appendix D). If the nucleus conforms to the J–J coupling scheme (and most do, at least approximately, and especially near closed shells) then the two states $J = l \pm \frac{1}{2}$ are distinct and orthogonal. They can be expressed in terms of orbital and spin product wave functions thus:

$$|l + \tfrac{1}{2}, m\rangle = \alpha \, |l, m + \tfrac{1}{2}\rangle \, |\tfrac{1}{2}, -\tfrac{1}{2}\rangle + \beta \, |l, m - \tfrac{1}{2}\rangle \, |\tfrac{1}{2}, +\tfrac{1}{2}\rangle,$$

$$|l - \tfrac{1}{2}, m\rangle = -\beta \, |l, m + \tfrac{1}{2}\rangle \, |\tfrac{1}{2}, -\tfrac{1}{2}\rangle + \alpha \, |l, m - \tfrac{1}{2}\rangle \, |\tfrac{1}{2}, +\tfrac{1}{2}\rangle,$$

where α and β are normalized such that $|\alpha|^2 + |\beta|^2 = 1$ (they are known, but the values are not needed here). Since $s_z = \frac{1}{2}\hbar \begin{pmatrix} 1 & 0 \\ 0 & -1 \end{pmatrix}$ it follows that $s_z |l + \frac{1}{2}, m\rangle$ contains both $|l + \frac{1}{2}, m\rangle$ and $|l - \frac{1}{2}, m\rangle$ as also does $s_z |l - \frac{1}{2}, m\rangle$. In treating s_x and s_y it is simpler to recombine them to form $s^+ = s_x + is_y$ and $s^- = s_x - is_y$ with weighting factors which will depend upon B_{0x} and B_{0y}. Now, from Appendix D, $s^+ |l + \frac{1}{2}, m\rangle \propto |l, m + \frac{1}{2}\rangle \, |\frac{1}{2}, +\frac{1}{2}\rangle$, which is a linear combination of $|l + \frac{1}{2}, m + 1\rangle$ and $|l - \frac{1}{2}, m + 1\rangle$; and similarly for s^-. Thus the effect of $s \cdot \boldsymbol{B}_0$ is to mix the original state $|j, m\rangle$ into states belonging to both couplings of l and s and also to mix the orientations, $\Delta m = 0, \pm 1$. Thus the matrix element will be non-zero if the final state contains one or other of these components. A similar argument applies for the term in $\boldsymbol{l} \cdot \boldsymbol{B}_0$ for the proton. The selection rule for magnetic dipole single-particle transitions is

M1: $\Delta l = 0$, $\Delta m = 0, \pm 1$, $|\Delta j| = 0, 1$, no change of parity,

where $|\Delta j| = 1$ allows the alternative coupling scheme. In the general case it is necessary to decouple and recouple with the rest of the nucleus giving

M1: $\Delta J = 0, \pm 1$, no change of parity, $(0 \rightarrow 0$ excluded$)$.

The next step would be to expand the exponential again to give higher magnetic multipoles, but we shall leave it at this point, except for observing that the parity will alternate with the higher magnetic multipoles, as for the electric, and that the ratio between transition probabilities is again $\sim (kR_N)^2$ on going from one multipole to the next. It is interesting to note that, whilst E1 transitions are essentially from one shell to another, M1 and E2 transitions can be made between states with the same basic shell structure. In fact most low-lying nuclear states are different configurations of the same basic states so one encounters M1 and E2 transitions much more often than E1.

From the foregoing it is possible to make crude estimates of the ratios of these three main types of transition. E2 is seen to be down by a factor $(kR_N)^2 \sim 1/2000$ on E1 for a medium-weight nucleus, whilst

$$(M1/E1)^{1/2} \sim \hbar^{-1}\mu_N g_s B_0 \int \psi_m^* s_z \psi_n \, d\tau / eE_0 \int \psi_f^* z \psi_i \, d\tau$$

$$\sim (\mu_N g_s / 2eR_N)(B_0/E_0)$$

$$= (\hbar/McR_N)(g_s/4),$$

where the integrals have been given the approximate values $\frac{1}{2}\hbar$ and R_N. From the uncertainty principle $\hbar/R_N \sim Mv$ leading to a ratio of intensities $\sim v^2/c^2$ which, for a typical kinetic energy inside the nucleus of order 50 MeV, takes the value $\sim \frac{1}{10}$. This is perhaps an overestimate, a factor $\sim \frac{1}{100}$ is closer.

Blatt and Weisskopf (1952) make estimates of the matrix elements, namely $\frac{3}{4}R_N$ for E1 and $\frac{3}{5}R_N^2$ for E2, based on the simple assumption that the radial wave functions are constant within the sphere of radius R_N and zero outside it for both initial and final states. For M1 they take $M1/E1 \sim 10(\hbar/McR_N)^2$, where the factor 10 allows for the large-spin g-factors of proton and neutron. Inserting these estimates and the values of the physical constants gives

$$\Gamma_\gamma(E1) = 0.07 E_\gamma^3 A^{2/3}$$

$$\Gamma_\gamma(E2) = 4.9 \times 10^{-8} E_\gamma^5 A^{4/3}, \qquad (5.2)$$

$$\Gamma_\gamma(M1) = 0.021 E_\gamma^3,$$

where Γ_γ is in eV, E_γ in MeV, and A is the mass number. The radiation width Γ_γ is connected with the decay half-life by $t_{1/2}(s) \sim 4.6 \times 10^{-16}/\Gamma_\gamma$ (see also Fig. 5.1).

Thus it is seen that the single-particle estimates lead one to expect E1 transitions to be stronger than M1 or E2 at ~ 1 MeV. But, as remarked previously transitions between single-particle states will rarely be found at such low excitation. In addition, there appears to be a tendency for neutrons and protons to move together in nuclei, and, from the remarks below concerning

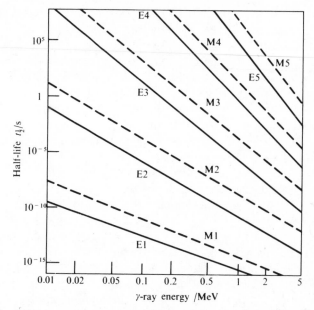

Fig. 5.1. Lifetime-energy relationships for γ-radiation, based on Weisskopf's single-particle estimates for $A \sim 100$.

centre-of-mass effects, this will very much reduce the effective charge of the radiating system for E1 transitions. Except for a few transitions in the light elements low-energy E1 transitions are many orders of magnitude weaker than single particle. Since M1 transitions are expected between similar components in complex wave functions, they are rarely found to approach single-particle strength, and are typically a few per cent of this value. On the other hand, E2 transitions are often found to be a factor of 100 stronger than the single-particle estimate. The matrix elements so far discussed have been treated implicitly in terms of the shell model in which only a small number (~ 1) of particles can find suitable initial and final configurations. If, however, all Z protons move coherently, then it is possible in principle to obtain transitions up to Z^2 stronger than single particle, since the power radiated depends upon the square of the charge of the effective particle radiating. In a rotational band the states have such a correlated motion in which about 20 nucleons appear to be taking part.

The switch from protons to nucleons is deliberate; it is found that neutron transitions are as strong as proton transitions for both E1 and E2, though for different reasons. In the former case it is because of the finite mass of the nucleus; when a proton moves outward, the rest of the nucleus must move in the opposite direction to keep the centre of mass unchanged. By simply taking first moments (in the classical sense) about the centre of mass it is seen that the system behaves as though it were infinitely massive provided that the charge on the proton is taken as $+e(1 - Z/A) \sim +\frac{1}{2}e$. A similar argument gives the effective charge on the neutron as $-eZ/A \sim -\frac{1}{2}e$ (see Appendix F). Since first moments were taken this applies to E1 transitions. Performing the same operations with second moments gives only small departures from $+e$ and zero for the two nucleons. Thus the centre-of-mass effect does not produce large E2 neutron transitions. That E2 neutron transitions are as large as proton transitions, even in nuclei like ^{17}O and ^{17}F (closed shell plus one neutron or proton), must be ascribed to a collective polarization of the core, and in both cases it must be a large effect.

Since E2 are in general larger than single particle and M1 in general smaller, they are found to be comparable in strength. The commonest transitions in nuclei are therefore found to be M1 or E2 or mixed M1/E2 transitions; in the latter case either type can be larger though with a tendency towards E2 in the heavier, distorted nuclei and to M1 in the lighter nuclei or nuclei near to closed shells.

Internal conversion

It is possible for an excited nucleus in an atom to de-excite by emitting an atomic electron by an electromagnetic process known as internal conversion. It has an atomic counterpart known as the Auger effect when, as an alternative to X-radiation, an outer shell electron is emitted instead. This process could be looked upon merely as the photoelectric effect in which the atom happens to be the same one as contains the radiating nucleus; indeed, such a process must occur and is more likely to occur for this particular atom than any other atom, since the radiation density is vastly greater than for any other atom. However, the photoelectric effect is not very probable for 'any other atom', so even for the particular atom it is likely to account only for a very small fraction of decays. But internal conversion can account for the major fraction of low-energy transitions in heavy nuclei, so a single-stage process is indicated in which the energy is transferred directly from the nucleus to the ejected electron. The Coulomb field provides the perturbation for the process.

The Coulomb potential at a given point r_e in space due to the nucleus is of the form

$$V(r_e) = \sum_i \frac{e}{|r_e - r_i|} \frac{1}{4\pi\varepsilon_0},$$

where the summation is taken over the protons, of position vector r_i, inside the nucleus. If the field point r_e lies outside the nucleus, i.e. $|r_e| > |r_i|$ for any r_i inside the nucleus, then the term $1/|r_e - r_i|$ can be expanded in powers of $r_i \cdot r_e / r_e^2$. When r_i is averaged over the nucleus each term in the expansion produces a nuclear electric multipole moment just as in the corresponding expansion of the radiation field. There is also a similar expansion in terms of magnetic multipoles arising from the magnetic interaction between the nuclear magnetic moment and that of the electron. Thus γ-emission and internal conversion depend upon the same nuclear property.

Without going into too much detail, the initial state is $\psi_0(\pi a^3)^{-\frac{1}{2}} \exp(-r_e/a)$ where a is the Bohr radius of the K-orbit for element Z and ψ_0 its nuclear initial (excited) state; the final state is $\psi_f \Omega^{-\frac{1}{2}} \exp(-i k_e \cdot r_e)$ where Ω is the volume of the arbitrarily imposed enclosure, k_e the wavenumber (and direction) of the emitted electron and ψ_f the final nuclear state (which may be a lower excited state or the ground state); and the perturbation is $-e$ times the potential given above. The matrix element for the (electric) process is therefore given by the integral

$$\frac{1}{(\pi a^3 \Omega)^{\frac{1}{2}}} \iint e^{-ik_e \cdot r_e} \psi_f^* \sum_i \frac{1}{4\pi\varepsilon_0} \frac{-e^2}{|r_e - r_i|} \psi_0 e^{-r_e/a} \, d\tau_N \, d\tau_e.$$

Evaluating this integral requires mathematical techniques beyond the standards expected of the reader of this book, but the following points can be appreciated:

(1) The terms in the expansion of the perturbation, as

indicated, generate the different dynamic multipole moments of the nucleus on performing the nuclear integration.

(2) When the transition energy exceeds the binding energy of the K-electron by more than a few eV then $k_e a \gg 1$ and under these conditions the rapid change of phase of the first exponential is far more effective in limiting the value of the integral than is the second exponential (as an example

$$\int_0^\infty e^{-(ik+\frac{1}{a})r}\, dr = \frac{1}{ik + 1/a} \sim 1/ik$$

for $ka \gg 1$). This second exponential can therefore be given the value unity with little error.

(3) Under the condition of (2), the integral is independent of a so the matrix element contains the factor $a^{-\frac{3}{2}}$ and the rate the factor a^{-3} i.e. $\propto Z^3$.

(4) Again under the conditions of (2), the scale factor of length inside the integral over the electron coordinates is $1/k_e$; since the length dependent terms come in, dimensionally, as $r_e^2\, dr_e/r_e^{(l+1)}$ for the lth multipole giving the electron integral the length dimensions of $(L)^{-(l-2)}$, this integral will therefore be proportional to $k_e^{(l-2)}$.

(5) The phase space factor is proportional to

$$\Omega p_e^2 \frac{dp_e}{dE_e}\, dE_e = \Omega(p_e^2/v_e)\, dE_e$$

giving $\Omega m_e p_e\, dE_e$ in the non-relativistic regime. The factor Ω eliminates this arbitrary volume from the final rate and the factor $p_e = \hbar k_e$ leads to an overall rate dependence $\propto Z^3 k_e^{(2l-3)}$.

The internal conversion coefficient α is defined as the ratio of probabilities of the conversion process to γ-emission. It is also further divided into components, $\alpha = \alpha_K + \alpha_L + \alpha_M$ etc., corresponding to the shell from which the electron is emitted. From the above considerations $\alpha(l+1)/\alpha(l)$ should be $\sim k_e^2/k_\gamma^2 = p_e^2/p_\gamma^2 \sim 2m_e c^2/\hbar\omega$ when E_γ is fairly large compared with the binding energy of the K electron, but the kinetic energy of the electron is small cf $m_e c^2$; for $\hbar\omega = 0.1\,\text{MeV}$ on a light nucleus ($B_K \ll 0.1\,\text{MeV}$) this has a value ~ 10. Thus the internal conversion coefficient increases strongly with Z for a given multipole order, and less strongly with multipole order for a given Z. Since α contains the factors $p_e^{(2l-3)}/E_\gamma^{(2l+1)}$ i.e. is proportional to

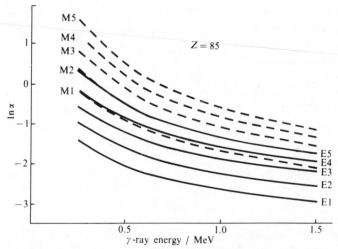

Fig. 5.2. Internal conversion coefficients (α) as functions of energy and multipolarity for a heavy element. (Based on Segre (1959). *Experimental nuclear physics*, Vol. III, Wiley.)

$(\hbar\omega - B_K)^{(l-3/2)}/(\hbar\omega)^{(2l+1)}$, it is at its maximum value at threshold (for $l = 1$) or slightly above (for $l > 1$) then falls off with increasing $\hbar\omega$. Similar considerations hold for the magnetic multipoles.

The behaviour of α is shown graphically in Fig. 5.2. The coefficients α_K, α_L, α_M, etc. have been extensively tabulated (in fact the α_L, α_M have been further subdivided, the former corresponding to the electron states $s_{\frac{1}{2}}$, $p_{\frac{1}{2}}$ and $p_{\frac{3}{2}}$) for all atoms over a wide range of energies and multipolarities. Measurements of α, α_K, α_L, α_M are used to determine the type of transition and even the amount of mixing in a mixed transition. Since the process is one of two-body breakup, the energy spectrum consists of discrete lines separated by the binding-energy differences of the atomic shells. Such lines were first discovered when superimposed on continuous β-spectra and initially led to confusion in the interpretation of β-spectra.

$J = 0 \rightarrow 0$ Transitions

It is not strictly correct to say that the conversion process has exactly the same nuclear dependence as that of γ-emission. The statement arose because of the similarity of two expansions and the expansion of $1/|\mathbf{r}_e - \mathbf{r}_i|$ is of the stated form only if $\mathbf{r}_e$ lies outside the nucleus. Whilst an atomic electron is usually well outside the nucleus, it does overlap it for a small fraction of the time; when $|r_e| < |r_i|$ the expansion becomes quite different, and in particular the monopole component, which for an external point gives a time-independent field because centre of charge and centre of mass coincide, can now have a time variation. Looked at simply, a monopole oscillation of the nucleus is a pulsating mode, having no effect on an external point, but for an internal point the potential depends upon the instantaneous nuclear radius and this is oscillating.

Although the finite size effect is small for transitions capable of radiating, for $0 \to 0$ transitions it is crucial. There are examples of such transitions of less than 1 MeV which cannot occur by any other process—the simultaneous emission of two γ-rays is possible but has never been detected. An example is the first excited state of ^{72}Ge $(0^+, 690\,\text{keV})$ which decays in this way with a half-life of 0.3 μs. Note that the transitions should be $0^+ \to 0^+$ (or $0^- \to 0^-$, though none of this type is known) i.e. there is no change of parity. If a 0^- state were found to be the first excited state in an even–even nucleus at an excitation of less than 1 MeV, it would decay by a second order process like double γ-emission or the simultaneous emission of a conversion electron and a γ-ray.

Above 1 MeV the process in the next section become possible.

Internal pair creation

In this process a state of excitation greater than $2m_e c^2$ (1.02) MeV) can lose energy by the creation and emission of an $e^+ - e^-$ pair. Since the emission can take place from inside the nucleus, $0 \to 0$ transitions can occur this way. Also the density of final states will increase more rapidly with energy for this three-body break-up than for internal conversion, so at high enough energies this process will dominate in $0 \to 0$ transitions. An example is the first excited state of ^{16}O $(0^+, 6.06\,\text{MeV})$ which decays by pair-emission in 5×10^{-11} s. Pair-emission in competition with γ-emission has also been observed. Again in principle it should dominate at high energies, from phase-space arguments, but long before it does so the levels have become virtual with many channels open for the emission of nucleons.

Problems

5.1. Estimate the lifetime of the first excited state of a (spinless) proton in a cubic box of side 5 fm. (Consult Pauling and Wilson (1935) for the quantum-mechanical problem.)

5.2*. What is meant by multipolarity of electromagnetic radiation? Explain why multipolarities other than E1 are commonly observed in nuclei but only rarely in atoms.

The decay scheme of low-lying levels in ^{14}N is (A; 0; 1$^+$, 0; $-$),   (B; 2.3; 0$^+$, 1; A, 100   per   cent), (C; 3.9, 1$^+$, 0; B, 96 per cent; A, 4 per cent) (D; 4.9; 0$^-$, 0; A 100 per cent), (E; 5.1; 2$^-$, 0; A, 67 per cent; B, 33 per cent), where the capital letter designates a level, of energy in MeV given next, followed by J^π, T, and finally the γ-decay of the level in percentages to the levels below. Sketch the level scheme and the transitions. Explain the significance of J^π, T and give the possible and expected multipolarities of transitions. Comment on the absence of the transitions DC, DB, ED, and EC. What other decay process can occur?

5.3. $^{160}_{65}$Tb decays by β-emission to states of $^{160}_{66}$Dy. Two of the γ-rays resulting are associated with an internal conversion line spectrum as follows; 32.9, 78.0, 84.7, 86.1, 143.0, 188.6, 195.1, 196.5 keV. The K, L, M, N, edges in Dy are 53.8, 8.6, 1.9, and 0.4 keV respectively. Find the γ-energies, identifying each conversion line with the nuclear transition and the atomic shell from which it arises. (Ignore the splitting of the L,M,N levels.)

5.4. Determine the effective charges due to centre-of-mass effects for neutron and proton E1 and E2 transitions in $N = Z$ nuclei, using the argument outlined on p. 102.

5.5. From the result of Problem 5.4, expressing the dipole operator in $N = Z$ nuclei as $-e \sum_i \tau_z^{(i)} r_i$, show that no E1 transition is possible between $T = 0$ states. (Consider specifically the isospin component of the wavefunction of a proton and a neutron outside closed shells. Consult Appendix F if difficulties arise.) Return to Problem 5.2 to determine whether this new selection rule has relevance.

6. Nuclear reactions

Introduction

In the earlier chapters the decay of unstable ground states was considered, and in the previous chapter the decay of low-lying excited states. The scope is extended here to states of high enough excitation to decay by particle-emission and also to consideration in more detail of the setting up of the low-lying states which decay by γ-emission. The experimental technique is to bombard a suitable material (the target) with a beam of nuclear projectiles and to observe the reaction products, or to select them for observation using coincidence techniques. If the projectiles are charged particles (p, d, α, ^{12}C, ^{16}O, etc.) they can be produced in quantity in a well focused beam of well-defined energy from an accelerator, e.g. a cyclotron or an electrostatic generator. If the projectiles are neutrons, then they must first be produced by a nuclear reaction, selected in energy as well as possible, and collimated (rather than focused) onto the target; the products from the second nuclear reaction are then observed, against the background produced by neutrons from the first reaction emitted in directions other than that of the second target. A copious source of neutrons is a reactor (see Chapter 7), especially for neutrons of low energy (the keV and eV region); but for neutrons of energy a few hundred keV and greater it is preferable to produce them using a primary reaction induced by charged particles from an accelerator.

The study of nuclear reactions has three basic aims:

(1) to locate states in a nucleus;

(2) to determine nuclear properties of states;

(3) to study reaction mechanisms, i.e. nuclear dynamics as distinct from the (quasi-) static behaviour in (2).

Experimentally there is much overlap of the three aims, especially (1) and (2), since the location of states and decay mechanisms can often be determined in the same experiment. To locate states we need to be able to handle the kinematics of nuclear reactions; to determine their properties we need to look at the decay modes and measure their lifetimes. For (3), it is intended to limit the discussion to a brief description of the direct and compound nucleus reaction mechanisms.

The kinematics of nuclear reactions

A simple, two-particle nuclear reaction can be represented as

$$A + a \rightarrow B + b + Q,$$

where a is the incident projectile and A the target nucleus, at rest in the laboratory frame; b is a light particle emitted in the reaction and B the final nucleus (the distinction between light and heavy final product is one of convenience; it is not always possible to make it—an example being fission (see Chapter 7)). The 'heat of reaction' Q, as defined above, is positive if the kinetic energies of the final products exceed that of the initiating projectile. The extra energy comes from the internal potential energies of the particles, and these are reflected in their masses. Elastic scattering is a particularly simple example of a nuclear reaction, having $Q = 0$. In general, the final nucleus need not be in its ground state; an increased state of excitation results in a reduced Q-value. From the kinematics of the reaction the Q-value is determined and so the state of excitation of B, provided the ground-state transition can be recognized.

If the reaction is represented in the laboratory frame as in Fig. 6.1, then using non-relativistic mechanics (accurate enough for most reactions using electrostatic generators and small cyclotrons): conservation of momentum:

$$\sqrt{(2m_a E_a)} = \sqrt{(2m_b E_b)}\cos \theta + \sqrt{(2m_B E_B)}\cos \phi$$
$$0 = \sqrt{(2m_b E_b)}\sin \theta - \sqrt{(2m_B E_B)}\sin \phi,$$

which on re-ordering, squaring, and adding give

$$m_B E_B = m_a E_a + m_b E_b - 2\sqrt{(m_a m_b E_a E_b)}\cos \theta,$$

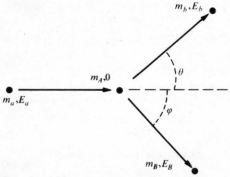

Fig. 6.1. Schematic representation of the kinetics of the reaction $A + a \rightarrow B + b$.

conservation of (kinetic) energy:

$$E_a + Q = E_b + E_B,$$

giving

$$(m_B + m_b)E_b - (m_B - m_a)E_a - 2\sqrt{(m_a m_b E_a E_b)}\cos\theta - m_B Q = 0,$$

$$(6.1)$$

conservation of (mass)† energy:

$$m_A + m_a = m_B + m_b + Q/c^2. \qquad (6.2)$$

Note that in the non-relativistic limit it is convenient to express the relativistic conservation of mass-energy as two separate equations connected by the heat of reaction, for a reason which becomes apparent in the next paragraph.

In the general experiment, we measure the energy of emission of b at a given angle of emission θ relative to the beam. In eqns (6.1) and (6.2) it is assumed that m_a, m_A, m_b, E_b, and θ are known (given data or by measurement) and m_B and Q are unknown. Thus (6.1) and (6.2) can be used to determine m_B and Q. In point of fact it is rarely necessary to solve these cumbersome simultaneous equations in the normal way. From

† The equation is given in terms of bare nuclear masses (m) whereas data tables give atomic masses (M). Since charge is conserved in nuclear reactions, the equation will also hold with atomic masses, provided that all masses are taken from data tables—the masses for p and α will then be listed as for the atoms ^{1}H and ^{4}He.

examination of the mass values of the stable and nearly stable nuclei it will be seen that the error incurred by using the mass number A for the mass of an unknown nucleus (and A will be known by conservation of nucleons) in (6.1) is probably of the order of 0.1 per cent, and certainly not greater than 0.3 per cent for all but the very lightest nuclei. It is therefore possible to solve (6.1) directly for Q to an accuracy ~ 0.2 per cent; since in (6.2) the term in Q is very small, the error incurred in deducing an accurate mass value m_B from (6.2) using this inaccurate Q is probably less than the errors arising from the quoted mass values of the other particles. In any case, if it is justifiable to work to greater accuracy, relativistic effects should be considered.

If we have measured a number of different energies of particle b at a given θ, corresponding to different energy levels in the final nucleus B (assuming that b itself has no excited state—true for n, p, d, α), then to each energy E_i can be assigned a Q-value Q_i, where $Q_0 - Q_i$ is the energy of the level in nucleus B. Q_0, the Q-value of the transition to ground, is either known or recognizable from a previously known pattern of excited states. Often the largest Q-value is assumed to arise from the transition to ground—occasionally an inaccurate assumption.

Often eqn (6.1) is used in reverse. In a given spectrum we may suspect that there may be spurious lines due to impurities in the target. We therefore wish to determine energies corresponding to levels in the final nucleus following reaction with the target impurity, for which Q_0 is known and the energy levels of the final nucleus. Thus in (6.1) all is known except E_b and this must be solved either as a quadratic in $(E_b)^{\frac{1}{2}}$ or by iteration.

Finally, in measuring differential cross-sections, it must be appreciated that the correct coordinate system to use is the centre-of-mass frame, not the laboratory frame. This affects the measurements in that the transformation from the latter to the former frame changes both the angle and the effective solid angle of the detector. The calculation has been left as an exercise.

Properties of states

Having located (bound) states in a nucleus using the above kinematic analysis, it is desired to determine other characteristics of these states such as spin, parity, γ-decay spectrum to lower-lying states, half-life for γ-decay, magnetic moment, electric quadrupole moment, and less obvious characteristics such as the extent to which a particular state looks like a single particle (proton or neutron) outside a core which could be the ground state (or an excited state) of the appropriate neighbour.

A great deal of information can be obtained by detecting the γ-radiation from the state, either directly or in coincidence with the particle which resulted in the formation of the state. The relative intensities of γ-transitions to known levels can give useful information on application of the selection rules. By the method of 'delayed coincidences' it is possible to plot the decay curve of the state and so determine its half-life. Used in conjunction with Weisskopf estimates (see p. 100), it may then be possible to determine the spin and parity of the state. A useful technique is to measure the angular correlation of the γ-radiation either relative to the beam, or with the outgoing particle detected in a certain direction (a most convenient direction is along, or against, the beam, since the problem then has axial symmetry)—see Fig. 6.2.

In order to see how information can be obtained from the direction of emission of particles from the nucleus, consider an idealized nuclear state which looks like an α-particle in an $l = 1$ orbit outside a closed shell core. Then, relative to a given axis, the angular components of the wave function will be of the form $Y_{1,m}(\theta, \phi)$. If the reaction setting up the state was selective in m

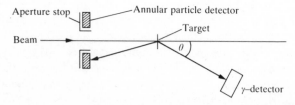

Fig. 6.2. A typical representation of an angular correlation experiment. For the reaction $A + a \rightarrow B^* + b$ followed by $B^* \rightarrow B + \gamma$, the emitted particle (b) is detected close to 180° by the annular detector through the centre of which the beam of particles (a) passes; the subsequent γ-ray is detected, in coincidence with a particle (b), at an angle θ with the beam axis. An additional relevant measurement with the above arrangement is to measure the θ-dependence of the γ-ray yield with particle (b) undetected—often referred to as the angular distribution of the γ-radiation.

and ensured that it was in the $m = 0$ sub-state then the angular wave function is $Y_{1,0}(\theta, \phi) \propto \cos \theta$; if also this state is unstable to the emission of α-particles, then the wave function of the free α-particle must have the same angular dependence (a necessary condition for matching wave functions at a spherical boundary). So the probability of observing an α-particle at an angle θ to the given axis will in this simple case be proportional to $\cos^2\theta$. It is unfortunate that the example chosen is that of an unbound state when discussing the γ-decay of particle-stable energy levels, but the object was to avoid the additional complexities of spin when considering effects of orbital angular momentum and this is not possible for γ-transitions (nor nucleon emission). Angular dependence is, however, implicit in the discussion of γ-emission in the previous chapter, where it was indicated that γ-absorption by a $(J = 0, M = 0)$ orbital state of a proton to form a $(1, 0)$ orbital state arose from the z-component of the radiation field. Since the field is perpendicular to the propagation vector, the z-component will contain the factor $\sin \theta$, where θ is the angle between the propagation vector and the z-axis. Thus the probability of absorption is proportional to $\sin^2\theta$ and, by time-reversal, the probability of decay of a $(1, 0)$ orbital state to a $(0, 0)$ orbital state by emission of a photon in a direction θ to the axis of quantization is proportional to $\sin^2\theta$. Another way of presenting this is merely to state the classical result that a dipole along the z-axis emits a radiation field whose amplitude is proportional to $\sin \theta$.

Thus, in general, if a level can be excited in a single sub-state relative to a given axis then the decay products (whether particle or γ-ray) will be emitted with an angular variation relative to that axis. A corollary to this statement is that if the level has equal population of all sub-states then the radiation will be isotropic, because lack of knowledge concerning the formation of the state can only be expressed in this way. It is rarely possible to set up a decaying level as a single sub-state, but neither is it necessary to do so to observe an angular dependence of the decay; a non-random population of the substates is sufficient, and this occurs whenever the particle initiating the reaction is absorbed out of a directional beam. Classically, the absorption of a particle moving in a certain direction cannot change the component of orbital angular momentum of the system along that direction and this holds good in making the transition to quantum mechanics. It is therefore convenient to make the beam direction the axis of quantization; the setting up of the state by absorption of a particle out of the beam is then characterized by $\Delta m = 0$. Thus if an α-particle were to be absorbed by a 0^+ target to form a 1^- state then this state could only be formed as the $(1, 0)$ sub-state, and its subsequent decay by γ-emission to a 0^+ ground state will give an angular distribution proportional to $\sin^2\theta$. If the target ground state has angular momentum J and the α-absorption produces a state $J' = J + 1$ then the substate (J, M) will result in the formation of $(J + 1, M)$. Since nothing is known about the target condition the angular distribution arising from each initial (J, M) must be summed with equal weighting. But in the inter-mediate state, the sub-states $(J + 1, J + 1)$ and $(J + 1, -J - 1)$ cannot be formed at all, so the averaging in the initial state is not equivalent to random averaging in the intermediate state and is not likely to produce isotropy. To proceed further requires a detailed study of the mathematics of coupling of angular momentum in quantized systems; here it suffices to state that an angular distribution can be calculated in terms of a few nuclear parameters, the J^π of the states involved, the mixing of emitted radiations (e.g. M1 and E2), etc., and checked against measure-ment to give information concerning these parameters.

Three important qualitative features can be appreciated. If the intermediate state has $J = 0$ then any radiation subsequently emitted must be isotropic—since randomness is represented by equal populations of all sub-states, $J = 0$ must always be in a random condition. Less obviously, if the particle absorbed has $l = 0$, then all subsequent radiations will be isotropic—this follows from the realization that adding nothing (in terms of angular momentum) to randomness can hardly produce other than randomness. Still less obviously, if the intermediate state has $J = \frac{1}{2}$ then any radiation subsequently emitted must be isotropic. This demonstrates the influence parity conservation has on the process: if one has a single substate (J, m) radiating and reflects the system in the x–y plane (from an experimental point of view this is equivalent to reversing the beam direction), then parity conservation requires that the system be physically the same. Since angular momentum is an axial vector it remains unchanged, $(J, m) \rightarrow (J, m)$; but relative to the transformed axes the substate has become $(J, -m)$ so parity conservation requires that the angular distribution from the substate (J, m) be identical with that from $(J, -m)$. To return to $J = \frac{1}{2}$, since $(\frac{1}{2}, \frac{1}{2})$ and $(\frac{1}{2}, -\frac{1}{2})$ have the same angular distributions on decay and, further, equal quantities of them give isotropy, then each has an isotropic distribution.

The wave function of a particle (of zero spin) being emitted in a state of well-defined orbital angular momentum l can be represented by the spherical harmonic $Y_{l,m}(\theta, \varphi)$. It transforms by the parity operation to $Y_{l,m}(\pi - \theta, \phi + \pi)$ which from the functional form of the harmonics, equals $(-)^l Y_{l,m}(\theta, \varphi)$. Thus even l = even parity, and odd l = odd parity. The associated intensity of this l-wave is represented by $Y_{l,m}^*(\theta, \varphi) Y_{l,m}(\theta, \varphi)$ which is independent of φ (since the two terms have the same m), and since it also must have even parity $((-)^{2l} = +)$ then it must be the same for θ and $(\pi - \theta)$, i.e. it has symmetry relative to the direction of the beam—a symmetric angular distribution. Now assume that the particle is in a mixed state, then on forming the intensity terms like $Y_{l,m}^*(\theta, \varphi) Y_{l',m}(\theta, \varphi)$ arise. Since the m values are the same (m is always well defined when the axis is the beam direction and the process starts and finishes in well defined substates), there is still rotational symmetry about the beam (independence on φ) and each particular term arising will be symmetric or antisymmetric with respect to the beam according as $l + l'$ is even or odd. Thus for mixtures of l-values all even or all odd the angular distribution is still symmetric, but for both even and odd l-values present this will not be the case (as in optics, the square of the modulus being positive definite, one cannot end up with a negative intensity). Going back a stage further to the intermediate state which must be a mixture of J^π states, then if all have the same π, the distribution will be symmetric, but if both π are present it will not be so. Inclusion of spin for the emitted particle clouds this simple picture, but does not affect the final result (assuming that polarization measurements are not being made).

The above outlines the theme, but the variations are numerous and can be complicated. As an example of complication, if a reaction is observed in the presence of a magnetic field then, from Larmor's theorem, the whole system can be assumed to rotate with the Larmor precession frequency; measuring the angular distribution as a function of time after formation of the intermediate state should reveal a rotation of the angular distribution by an amount which depends upon the magnetic moment of the intermediate state and the time delay and so gives the magnetic moment of this excited state.

The methods for determining to what extent a given state looks like a shell-model state are best left until the next section.

Interaction mechanisms

Broadly speaking interactions can be divided into two categories, but the division is not sharp. The category of 'direct reactions' refers to reactions which take place whilst the bombarding particle is traversing the target nucleus, i.e. in a time $\sim 2R_N/v$, where v is the velocity of the particle (typically $>c/10$). This characteristic time is 10^{-21}–10^{-22} s. An example is the reaction $A(d, p)B$, in which the deuteron, whilst passing through the periphery of nucleus A loses a neutron to form B. In the second category, 'compound-nucleus reactions', the time of interaction is very long compared with the characteristic time above. Thus there is a long time interval between the absorption of the bombarding particle and the emission of the product particle. It is therefore reasonable to talk of the compound nucleus as an intermediate product in the reaction; if the reaction is $A(a, b)B$ then it effectively proceeds in two steps $A + a \rightarrow C^* \rightarrow B + b$. Since C^* exists for a time long on the nuclear time scale it must correspond to a (quasi-) stationary state of the nucleus C and is in fact just one of the many virtual states of excitation of C.

As is usual, we define the processes in terms of extremes. The real situation is not always so clear. If an incoming projectile interacts with a nucleon near the surface of the nucleus, it or the nucleon could be scattered clear of the nucleus giving a fast direct reaction; if it gives up some energy to the nucleon but insufficient to raise the latter out of the nucleus and thereby moves further into the nucleus, it can once again be scattered out by another nucleon or share energy with it and still remain inside the nucleus. At each collision it becomes harder for the particles concerned to escape, and finally the energy is well shared among all the nucleons. At this stage the compound-nuclear state has indeed been formed. It can only decay when, by chance, one nucleon can acquire a sufficiently large share of the excess energy to become unbound, and this can take a long time when the energy has been shared over a large number of nucleons. Thus virtual nuclear states tend to be longer lived (for a given excitation) in heavier nuclei. At low energies, and especially for charged particles, the compound-nucleus process will dominate in the reaction cross-section (the elastic Rutherford scattering is large and direct) since, having penetrated the barrier (an improbable process), the particle even if it fails to collide has little chance of penetrating once again at the first pass. It is therefore reflected back and forth at the surface until finally it shares energy and forms the compound nucleus. At high energies direct reactions become of increasing importance—at very high energies it is virtually certain that both incoming and struck particle will escape from the nucleus.

As previously stated, the compound nuclear state C^* is just one of many virtual states in the nucleus C, each being capable of playing the rôle of compound nuclear state. At low energies of the incident particle the states in C are well separated leading to isolated resonances in the compound nuclear process, but as the energy is increased the states in C crowd closer together and at the same time get broader (though rather slowly—this process is discussed in Appendix G) and eventually overlap. The resonances in the compound nuclear process behave similarly and the cross-section for a given reaction at a given energy may receive contributions from a large number of unresolved resonances. In the following treatment it is assumed that the energy is low enough for the virtual states in C to be well separated, giving a simple isolated resonance at the appropriate energy range.

The compound-nucleus process, $A + a \rightarrow C^* \rightarrow B + b$

The reaction can be described mathematically as arising from a first-order time-dependent perturbation setting up the compound nucleus C^* which subsequently decays, or alternatively as arising from a second-order time-dependent perturbation between the initial and final states which is dominated by one channel proceeding through C^*. The two processes are in reality the same thing, but it is instructive to look at both. For a first look at the problem we will ignore the fact that each state may be degenerate (since J may not be zero and the particle absorbed or emitted may also have spin) and assume that states A, B and C* are unique, as also are the absorbed and emitted particle states.

As always in time-dependent perturbation, we consider the amplitudes a, b, c of normalized wave functions of states A, B, and C* to be functions of time. The wave function representing C* is taken to be of the form $c_0 \exp(-t/2\tau)\psi_C(r, \theta, \phi)$ in order to represent the fact that, left alone, C* will decay in intensity with mean life τ. First-order perturbation theory gives the rate of formation of C* as

$$-(i/\hbar)H_{CA} \exp\{i(E_C - E_A)t/\hbar\}a,$$

where H_{CA} will be assumed to be independent of time, except for having been switched on at $t = 0$. H_{CA} represents the spatial part of the complete matrix element of the perturbing potential V_{Aa} responsible for the process A + a → C*. It is independent of time because the incoming beam is represented as a plane wave, which corresponds to the kinetic energy, E_A in the CM coordinate system of the incident channel, being known precisely. Under these conditions the time dependence of the matrix element is just the exponential factor in the above expression, where E_C is the energy of the state C* in nucleus C measured relative to the threshold energy in the incident channel (alternatively E_A can be defined as the total energy—rest plus kinetic—in the CM coordinate system of the incident channel and E_c will then be the rest energy of nucleus C in the state C*). The amplitude c therefore satisfies the differential equation

$$\frac{dc}{dt} = -(1/2\tau)c - (i/\hbar)H_{CA} \exp\{i(E_C - E_A)t/\hbar\}a.$$

If the probability of transition is small, then there is little absorption of the wave representing the initial state, so $a = \text{constant} = 1$ is a reasonable approximation, and the equation integrates to

$$c = \frac{-(i/\hbar)H_{CA}[\exp\{i(E_C - E_A)t/\hbar\} - \exp(-t/2\tau)]}{(i/\hbar)(E_C - E_A) + 1/2\tau}, \quad (6.3)$$

satisfying the condition $c = 0$ at $t = 0$. While τ is large on a nuclear time scale it is small on a laboratory scale ($\sim 10^{-14}$ s or less), and so the term $\exp(-t/2\tau)$ can be neglected under experimental conditions giving

$$|c|^2 = \frac{|H_{CA}|^2}{\{(E_C - E_A)^2 + \hbar^2/4\tau^2\}} = \frac{|H_{CA}|^2}{\{(E_C - E_A)^2 + \Gamma^2/4\}}, \quad (6.4)$$

where $\Gamma\tau = \hbar$. Now the probability of decay of a state is the sum of its decay probabilities to the various decay channels:

$$1/\tau = \sum_i (1/\tau_i) \quad \text{or} \quad \Gamma = \sum_i \Gamma_i.$$

Rate of formation of state B $= \dfrac{|c|^2}{\tau_\beta} = \dfrac{|H_{CA}|^2\, \Gamma_B}{\hbar\{(E_C - E_A)^2 + \Gamma^2/4\}}.$

Considered as a second-order perturbation the rate of production of B is

$$(2\pi/\hbar)\left|\sum_c H_{CA}H_{BC}/(E_A - E_C)\right|^2 \times (\text{energy density of final states}).$$

The time-dependence of the (decaying) state C^* is

$$\exp(-iE_C t/\hbar)\exp(-t/2\tau) = \exp\{-i(E_C - i\Gamma/2)t/\hbar\}.$$

Thus the decaying state C^* is effectively at an energy $(E_C - i\Gamma/2)$. If in the above summation one term dominates and that one is C^*, then the expression inside the moduli becomes

$$\frac{|H_{CA}|^2\, |H_{BC}|^2}{\{(E_A - E_C)^2 + \Gamma^2/4\}}.$$

Comparison of the two expressions for rate of formation then gives

$$\Gamma_B = 2\pi\, |H_{BC}|^2 \times (\text{energy density of final states}),$$

and, by applying a time-reversal argument,

$$\Gamma_A = 2\pi\, |H_{CA}|^2 \times (\text{energy density of final states of}$$
$$\text{time-reversed system}).$$

From Chapter 4 (p. 81) this is

$$\Gamma_A = 2\pi\, |H_{CA}|^2 \left\{\frac{4\pi p_a^2}{v_a(2\pi\hbar)^3}\right\}$$

where the surrounding box has been taken to have unit volume (in any case the size drops out since it occurs also in the matrix element), and

$$\frac{\mathrm{d}n}{\mathrm{d}E} \propto p^2 \frac{\mathrm{d}p}{\mathrm{d}E} = p^2/v.$$

Thus

$$\Gamma_A = (1/\pi\hbar^3) |H_{CA}|^2 m_a^2 v_a, \tag{6.5}$$

and the rate of formation of B is then

$$\frac{\pi\hbar^2}{m_a^2 v_a} \cdot \frac{\Gamma_A \Gamma_B}{(E_A - E_C)^2 + \Gamma^2/4}.$$

From Appendix A, the cross-section for a process is defined such that $\sigma_{AB} v_a$ gives the rate of formation of state B. Therefore

$$\sigma_{AB} = \frac{\pi\hbar^2}{m_a^2 v_a^2} \frac{\Gamma_A \Gamma_B}{(E_A - E_C)^2 + \Gamma^2/4}, \tag{6.6}$$

and the factor $\pi\hbar^2/(m_a^2 v_a^2)$ can be written more familiarly as $\pi\lambda^2$. In this expression E_A is the energy of the incoming beam of particles (in the centre-of-mass frame) and is therefore the energy variable. The shape of the resonance curve is shown in Fig. 6.3, where the variations of $\pi\lambda^2$ and the widths over the resonance have been neglected.

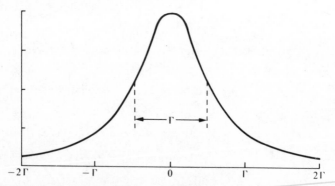

Fig. 6.3. Shape of Breit–Wigner resonance denominator plotted against $(E - E_r)$, with energy units of Γ. If the resonance is sharp so that neither λ^2 nor the widths vary appreciably over an energy range $\sim 4\Gamma$, then this curve will represent the behaviour of the reaction cross-section.

Now the restrictions on degeneracy will be removed. At this stage it is necessary to look a little closer at a point which was glossed over in the initial calculation. The final result refers to partial cross-sections, not differential cross-sections, but the wavefunctions for the states involved, including the ingoing and outgoing particles, have angular dependences based on some choice of axes to describe the system, which for convenience are chosen to be polar coordinates relative to the incoming beam direction. If now outgoing particles are observed in the direction (θ, φ) relative to this axis then the matrix elements should be functions of this direction, and to arrive at the partial cross-section integration over all (θ, φ) has implicitly been made. Also, now that degeneracy has been introduced, the matrix elements should specify the substates of A, B, and C and the spin components of the particles involved and appropriate summations made over these specifications; the appropriate summations may need to be made in amplitude or intensity (in amplitude for intermediate states and in intensity for initial and final states— this can be justified by imposing a very weak magnetic field on the system which can resolve the different substates of the initial and final states, which can in principle be long lived and therefore well defined in energy, but not the intermediate states which have a poorly defined energy arising from the short lifetime). However, on integrating over all directions of emission, interference terms drop out, so for integrated cross-sections all sums over substates are made in intensity. There is a classical analogy to this: a diffraction grating of N slits produces an angular distribution consisting of a small number of very large narrow peaks in intensity separated by a large number $(\sim N)$ of subsidiary peaks of much reduced intensity, but the intensity of light integrated over all angles is just N times the intensity falling on a single slit. Interference does not create intensity, it merely moves it around.

A 'proper' treatment of the problem would in addition introduce vector coupling coefficients which arise in quantum mechanics whenever angular momenta are combined, and the appropriate factor representing all the degeneracies would arise from theorems involving summing over the squares of such coefficients. In avoiding the use of these techniques some uncertainties arise—not concerning the initial and final states which can be dealt with reasonably convincingly, but rather with the intermediate state. To consider the initial and final states first, it is obvious that, for the second-order perturbation, the rate of production of B should include the factors $(2J_B + 1)(2J_b + 1)$ in the term (density of final states), and it is equally obvious that the same factors should arise in the same bracket in the expression for Γ_B so σ_{AB} requires no amendment for degeneracies in the outgoing channel. In the ingoing channel however, we have chosen to describe the process in terms of Γ_A which involves the outgoing channel in the time-reversed system and therefore contains the factors $(2J_A + 1)(2J_a + 1)$ built into it. The cross-section σ_{AB} must therefore be multiplied by $1/(2J_A + 1)(2J_a + 1)$ to remove these extraneous factors. One usually differentiates between initial and final states by saying that one sums over final states but averages over initial states—however many substates are available in the ingoing channel, it is specified by *one* particle moving towards the target nucleus. To cope with the intermediate state likewise, at first sight, seems the correct thing to do if we use the first-order calculation since, in it, C appears as a final state which happens to decay, and its density of states factor, which is unity with regard to phase-space factors since it is uniquely specified by the beam energy and direction, will acquire the factor $(2J_C + 1)$. This is quite correct, but mystifyingly so if one looks at it a little closer; specify J_A and J_a as both zero, then only the $m = 0$ substate of C can be produced by absorption of a particle moving along the axis of quantization. The summation over final states is not so obvious as it initially appeared to be. Looked at as a second-order perturbation we now have $(2J_C + 1)$ states of the same energy giving parallel paths along which the process proceeds. From what has been said about interference not changing the integrated intensity, and assuming all substates are equivalent, then a factor $(2J_C + 1)$ should again arise in σ_{AB}. But the considerations presented above still apply and, for $J_A = J_a = 0$, the summation would again collapse to one term. The proper treatment would take care of this, but can we get out of our dilemma with our simple approach? It arises because we have allowed experimental convenience to reduce a problem which has spherical symmetry to one of lower symmetry namely axial symmetry. It may be experimentally inconvenient to produce a beam converging isotropically onto the target, but mathematically it is easily visualized—quite obviously all directions of the incoming beam should give the same cross-section in the absence of polarization. With no axis specified by the problem the assumption that all substates are equivalent should now hold good and the factor $(2J_C + 1)$ should arise in σ_{AB}—it is the further effective integration over all angles of the incoming beam which results in equivalent weighting of all substates, whereas for the case of lesser symmetry it is not so obvious.

To proceed to the general expression

$$\sigma_{AB} = \pi \lambdabar^2 \frac{(2J_C + 1)}{(2J_A + 1)(2J_a + 1)} \frac{\Gamma_A \Gamma_B}{(E_A - E_C)^2 + \Gamma^2/4}.$$

In the derivation of the Breit–Wigner formula above it has been assumed that the decay of C* is characteristic of C* alone and does not depend upon the way C* was produced. This independence of formation and decay of the compound nucleus was first stressed by Bohr. It is an important point, but has been somewhat overworked in the textbooks. It holds only for an isolated resonance, but in tests quoted, which are at regions of high excitation in the compound nucleus, different formation processes can populate the overlapping resonances to different degrees; one reason for this is the dependence of the angular momentum barrier on the mass of the projectile, another is the randomness of level widths in nuclear states. The decay processes will then differ. If the number of overlapping resonances is very large then a statistical averaging process will reduce the difference. Thus independence of formation and decay is good for a single resonance or a large number of overlapping resonances, but probably does not hold for a few overlapping resonances.

Two outstanding examples of resonant reactions are shown in Figs. 6.4 and 6.5. In the (p, γ) reaction the resonances are sharp because the proton width is small due to the effect of the potential barrier. From the above, Γ is proportional to the square of the modulus of a matrix element which connects a state representing a particle within the nucleus with a state representing the same particle outside the nucleus (a plane wave at infinity) and the overlap of this latter wave function with the nucleus introduces the barrier-penetrability term. It is then

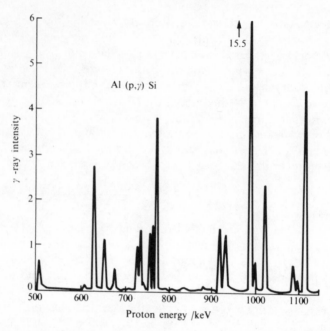

Fig. 6.4. Resonances in ^{27}Al(p, γ)^{28}Si. The resonance widths indicated represent experimental limitations; the true widths are considerably less. (Based on Brostrom *et al.* (1947). *Phys. Rev.* **71**, 661.)

surprising that the neutron resonances shown have even smaller neutron widths. This is partly accounted for by the fact that the separation between resonances has become small because in the examples shown the target nucleus in (n, γ) is a heavy one and in (p, γ) a light one. But why should crowding of levels result in smaller partial widths? This question can be conveniently answered by considering what would happen to cross-sections were it not the case. Assume that the cross-section is being averaged over an energy range ΔE large compared with resonance widths, then the contribution to the *total* reaction cross-section of a single resonance is obtained by integrating the Breit–Wigner expression (with Γ for Γ_B) giving a result proportional to Γ_A only. The number of levels in ΔE is $\sim \Delta E/D$, where D is the average spacing between levels. The averaged cross-section is therefore proportional to Γ_A/D. As the energy is increased D will get smaller corresponding to an increase in level

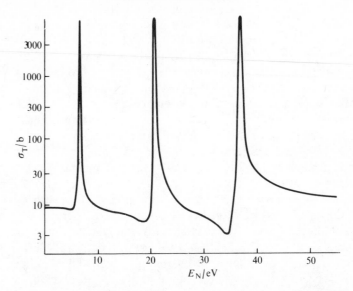

Fig. 6.5.   Resonances in ^{238}U(n, γ)^{239}U for neutrons up to 50 eV. Note the large cross-sections at resonance, the high density of levels in the compound nucleus (~1 every 15 eV) and the narrow widths. Note also that the measurements were made in absorption. Off-peak the absorption is due to 'hard-sphere' scattering interfering with the scattering from tails of the resonances; on-peak the absorption is dominated by the capture process.

density with excitation of nucleus C. If Γ_A remains approximately constant with energy then the cross-section increases without limit. But cross-sections at high energies should approach something like the classical πR_N^2, and we are forced to conclude that (Γ_A/D) is roughly independent of energy. Thus the partial width into a given channel depends upon the local spacing of levels (see also Chapter 3 and Appendix G). This does not mean that the total width of the resonance is similarly limited; it is a sum of all partial widths and each such partial width is proportional to the local spacing, but, as the energy increases, the number of open channels contributing also increases. Thus the total width of levels tends to increase with energy, but only slowly since the two effects almost compensate.

A reasonable statistical theory of nuclear reactions came on the scene rather later in the day (late fifties and early sixties). Before this it was thought that, as level densities got higher, the strongly fluctuating nuclear cross-sections (depending on whether one was on or off resonance) would quickly average out to smooth cross-sections as predicted by the optical model (see also Appendix G). But superposition of the random amplitudes and phases of overlapping resonances will always produce a wildly fluctuating intensity, even producing zeroes in intensity. This latter statement refers to measurements in one exit channel with an energy resolution (beam resolution plus target thickness) small compared with the widths of the overlapping resonances and refers to the differential cross-section (i.e. measurements at a specific angle relative to the beam). Any averaging process serves to damp out these fluctuations, e.g. integrating over a large angle of detection (to the extent perhaps of all angles), or over a large energy range (using a thicker target for charged projectiles), or combining the measurements in many exit channels (an example being measuring the neutron yield using a detector which does not depend on the neutron energy). The fluctuations in cross-sections form an interesting study. A local high in the cross-section cannot fade away over an energy range small compared with the local level width since the component parts of the amplitude change little in magnitude or phase over such small changes of energy. A suitable correlation analysis of cross-sections yields the local average level width (Γ_{tot}) and confirms the basic ideas of resonance cross-sections namely that, though levels crowd together very rapidly as the energy is increased, the average total width of a resonance changes very slowly over the same energy range, and this in turn indicates that the partial width into a single exit channel must indeed decrease rapidly with excitation, confirming the argument of the previous paragraph.

To return to (n, γ), it was stated that the above considerations only partly account for the small neutron widths. A closer look at the expression for the partial width in terms of the matrix element (p. 120) reveals that the width is proportional to the velocity in the channel and this is very small for neutrons of a few eV—so small that in this region $\Gamma_n < \Gamma_\gamma$. (This surprising circumstance is also a consequence of the fact that γ-radiation is taking place from a level at an excitation of ~6 MeV or more in a large nucleus, so there is an extremely large number of available states of lower energy to which γ-transitions can take place.) In addition, the width includes the penetrability through the angular momentum barrier (and for charged particles it is augmented by the Coulomb barrier) with the result that Γ_n in the eV region is far too small, when $l \neq 0$, to give observable resonances.

In the case of the (p, γ) resonances, their sharpness is determined largely by the barrier factor. The 'widths without barrier' in light nuclei are quite large fractions ($\sim$ a few per cent) of the single-particle proton width, but the barrier penetrability reduces them to a few eV giving a total width for the resonance less than the experimental resolution of the beam. The peak yield is not therefore the yield at resonance, but a measure of the integrated yield over the resonance.

Finally, the Breit–Wigner formula can be used to show that neutron cross-sections vary as $1/v$ at very low energies. So far it has been assumed that the Γs are constants in the formula but this is an oversimplification; the matrix elements can have an inbuilt energy dependence and, as has been mentioned, there is a factor v_a in the expression for Γ_a. For a sharp resonance at high energy in the incident channel the variation is not expected to be large over the region of large cross-section, but at very low energies the $\pi \lambdabar^2$ term becomes large and enhances the low energy tail of the resonance. It is therefore necessary to consider the contribution made by a resonance far from the peak of the resonance denominator. The cross-section is then of the form $\sigma \propto \pi \lambdabar^2 \Gamma_n \Gamma_\gamma / (E_C)^2$, assuming that $E \ll E_C$ and also $\Gamma \ll E_C$. Now Γ_γ is unlikely to vary whilst the energy changes by a few eV since the level in C is 6 MeV or higher in excitation, so the energy variation is contained in $\lambdabar^2 \Gamma_n$, and since $\lambdabar \propto 1/v_n$ and $\Gamma_n \propto v_n$ then $\sigma \propto 1/v_n$.

Direct reactions

The characteristic of a compound-nucleus resonance is a cross-section which changes rapidly with energy over a small range of energy. The range over which the cross-section changes rapidly is of order Γ and is correlated with the time scale of the process by $\Gamma \tau \sim \hbar$. A direct reaction has been defined as one taking place whilst the bombarding particle passes through the nucleus and therefore in a characteristic time $\sim 10^{-22}$ s. Using the above relationship, it is associated with a width ~ 10 MeV. Thus a characteristic of direct reactions is that the cross-section should not change rapidly with energy, or rather should not exhibit sharp maxima, since it is possible for a direct reaction involving charged particles below the barrier to increase exponentially in cross-section.

Another characteristic concerns the angular distribution of the process. If a reaction proceeds via a single (and this must be emphasized), well defined resonance then, at the intermediate stage of the process, it has a well defined parity (the parity of the state C*). In setting up the state a definite direction has been singled out (the direction of the beam), but this does not correspond to a well defined parity, so in forming C* the information concerning direction of motion is lost but not the axis of the beam. (To form a state of well defined parity it is necessary to superimpose upon the state defined by the beam, a corresponding state with the beam moving in the opposite direction.) Thus the subsequent decay of C* will give an angular distribution having symmetry about 90° (fore–aft symmetry). The direct reaction has no such restriction, the symmetry axis contains the information concerning direction and angular distributions do not in general have fore-aft symmetry. In fact there is a strong tendency to forward-peaking, an extreme example being the case of absorption of a fairly high-energy neutron beam which results in the diffractive scattering of the beam. As in the classical case of diffraction of light, the forward-directed scattering peak has angular width $\sim\lambda/R_N$ rad. Many other types of direct reaction conform to this simple diffractive picture.

The stripping reaction is an interesting example of a direct reaction. It is the reaction A(d, p)B, or equivalently (d, n), in which one nucleon of the incoming deuteron interacts with the nucleus whilst the other passes it by. The probability of the process occurring depends upon the square of the modulus of $\int \psi_f^* V \psi_i \, d\tau$, where the initial wave function ψ_i is that of the deuteron in the form of a plane wave combined with its internal function, multiplied by the target wave function ψ_A, whilst the final wave function ψ_f will be that of the proton in the form of a plane wave, at an angle θ relative to the axis i.e. to the deuteron beam, multiplied by the wave function of the final nuclear state ψ_B, which for simplicity will be taken as $\psi_A \psi_n$, where ψ_n is a single-particle wave function of the neutron about the core A. Notice that the Coulomb distortion of the plane waves has been neglected here, though it is included in the analysis of research data. For the perturbation take the n–p interaction, $V(|r_p - r_n|)$ —which can be justified by considering the inverse mechanism, the pick-up reaction, $B(p, d)A$ in which the proton may be considered to be pulling the neutron out of nucleus B. It is also assumed that any change of shape of the deuteron during its journey past the nucleus can be neglected and that reaction only occurs when the proton and neutron in the deuteron are at their common centre. Another simplifying assumption made here is that the reaction takes place only at the surface of nucleus A—this is not assumed in a more serious calculation.

With these approximations, in the integral, the internal wave-function of the deuteron becomes merely a constant (the probability that the proton and neutron should be found at its centre), and its plane wave becomes $\exp(i\boldsymbol{k}_d \cdot \boldsymbol{r}_n)$, whilst that of the proton becomes $\exp(i\boldsymbol{k}_p \cdot \boldsymbol{r}_n)$, since $\boldsymbol{r}_n = \boldsymbol{r}_p = \boldsymbol{r}_d$, by assumption. By the same token $V = V(0) = \text{constant}$, as also is $\psi_A^* \psi_A$. In the integral $\boldsymbol{r}_n \equiv (R_N, \theta', \phi')$ and the integral becomes

$$\text{constant} \times \int \psi_n^*(\boldsymbol{r}_n) \exp(i\boldsymbol{k}_d \cdot \boldsymbol{r}_n) \exp(-i\boldsymbol{k}_p \cdot \boldsymbol{r}_n) \, d\Omega'.$$

If $\boldsymbol{k}_d - \boldsymbol{k}_p = \boldsymbol{q}$, and the direction of $\boldsymbol{q}$ is taken to define the z'-axis the integral is proportional to $\int Y_{l,m}^*(\theta', \phi') \exp(iqR_N \cos \theta') \, d\Omega'$, where the single-particle angular wave function has been inserted for the neutron, but its spin function has been neglected. The integral over ϕ' requires that $m = 0$, giving

$$\int_{\cos\theta'=-1}^{\cos\theta'=+1} Y_{l,0}(\theta') \exp(iqR_N \cos \theta') \, d(\cos \theta').$$

To see how this behaves take the cases $l = 0$ and $l = 1$. The first integral, for $l = 0$, can be performed at sight, giving $\text{constant} \times (\sin qR_N / qR_N)$, which, when the substitution $q^2 = (k_d - k_p)^2 + 4k_d k_p \sin^2(\theta/2)$ is made, gives a maximum at $\theta = 0$, followed by subsidiary maxima of reduced intensity. (The function $(\sin qR_N / qR_N)^2$ is recognizable as the diffraction function, so, provided $|\boldsymbol{k}_d - \boldsymbol{k}_p| R_N$ is less than π, the statement made is true. This is the case in most experimental applications.) Putting in $Y_{1,0} \propto \cos \theta$, for $l = 1$, gives an integral $\int_{-1}^{1} \exp(iqR_N c) c \, dc$. Integrating by parts gives $(\sin qR_N - qR_N \cos qR_N)/(qR_N)^2$ which, on squaring, is seen to give zero at $q = 0$, rising to a large first maximum and then having subsequent lesser maxima. With the same proviso as for $l = 0$, the angular distribution will be non-zero at $\theta = 0$ (unless $k_d = k_p$ which is possible, for low E_d and a high Q-value, but not likely) rising to a large first maximum (at a fairly small angle in practical cases) followed by subsidiary maxima. The important point is that the location of the primary maximum can be used to determine the l-value of the transfer, i.e. the amount of orbital angular momentum taken in by the stripped nucleon. Thus the parity of the final state is known immediately on analysis of the angular distribution of the particle not captured, and the range of J-value of the final state is limited. The degree of limitation depends upon J for the initial nucleus—if this were zero then the final state must have $J_f = l_n \pm \frac{1}{2}$, where l_n is the transferred angular momentum. But in the general case J_f can take the range $(J_i + l_n + \frac{1}{2})$ down to the smaller of $|J_i - l_n \pm \frac{1}{2}|$. Measurements of proton polarization, brought about by the spin–orbit term of the optical potential, reduce these ambiguities and, for the particular case of a zero spin target nucleus, can

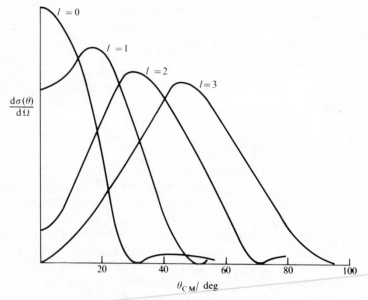

Fig. 6.6. Typical angular distributions for the (d, p) direct reaction. If each process represented the full single-particle transfer, then relative to $l = 0$, the curves for $l = 1, 2, 3$ should be multiplied respectively by $\frac{1}{4}, \frac{1}{12}, \frac{1}{30}$.

differentiate between $J_f = l_n \pm \frac{1}{2}$. Examples of angular distributions in (d, p) are given in Fig. 6.6.

A useful feature of the stripping reaction is that the magnitude of the cross-section is a measure of the degree to which the final state looks like the original state plus a single particle of the type indicated by the angular distribution. Of especial importance is the fact that the final state may be bound to the particle transferred (indeed the calculation is simpler for such bound states, and the simple assumptions made here would not be good for an unbound state). Because the transfer is virtual rather than real, the real positive momentum q can be associated with a negative total energy of transfer.

Apart from deuteron stripping, perhaps the most useful, and certainly the best understood, of the direct reactions is Coulomb excitation, especially with heavy ion projectiles. Place a charge at a distance R from a nucleus and the interaction can be expanded in terms of the multipole moments as given on p. 25. The dipole moment is down by a factor $\sim(R_N/R)$ on the monopole term since $(\partial U/\partial r)_0$ is down by $1/R$ on $U(0)$ and the dipole moment is of order eR_N; similarly the quadrupole term is down by the same factor on the dipole. Now allow R to be a function of time and one is perturbing the nucleus with time dependent multipole moments. In this particular case of a heavy ion passing a nucleus, the interaction is in the form of a short pulse in time which can be Fourier analysed into a range of frequencies. If this range of frequencies spans the absorption frequency for an internal excitation of the nucleus, then this internal excitation can occur. Since this excitation requires a certain change of angular momentum and parity of the nucleus, it is effectively instigated by the required multipole moment of the interaction. To be completely general we should have included magnetic terms in the interaction since the moving charge has the effects of a current, but we will dismiss them in detail by asserting, as perhaps expected, that a given magnetic multipole term will be down by a factor $\sim v/c$ in amplitude on the corresponding electric multipole term, where v is the velocity (at infinity) of the charged projectile.

The time of passage of the heavy ion past the nucleus is $\sim a/v$ where $2a$ is the closest distance of approach in head-on collision, and the Fourier component of angular frequency ω will be present in quantity provided $\omega a/v \lesssim 1$. This criterion leads to the result, perhaps obvious, that the probability that the process should occur increases as kinetic energy of the heavy ion increases (since $2a = Z_1 Z_2 e^2/4\pi\varepsilon_0 E_K$ where E_K is the kinetic energy in the centre of mass system). If one were to increase E_K without limit however the heavy ion will pass over the Coulomb barrier and disappear into the nucleus. This introduces both theoretical and experimental difficulties. Mathematically the process is no longer so clear in interpretation—the less well known nuclear interaction must now contribute to the potential complete with its imaginary (optical model) component to produce absorption whereas the simpler Coulomb term is sufficient at larger values of $2a$; experimentally the onset of nuclear reactions produces an enormous increase in unwanted radiation, making the process sought harder to detect. For these reasons the incident E_K is usually limited to perhaps 3/4 of the barrier height (the kinetic energy at which $2a = R_1 + R_2$ where R_1 and R_2 are the radii of nucleus and projectile respectively). Under these circumstances the effectiveness of the process does not vary greatly with multipole order, assuming that the multipole order λ has moment eR_N^λ, since $(R_N/R)^2$ (squared to go from amplitude to intensity) is as large as $\sim 1/10$ at the above limit. By contrast the ratio of effectiveness of multipoles in the de-excitation process by γ-emission is $\sim(R_N/\lambda)^2$ or $\sim 1/1000$ (see Chapter 5). The above estimates are based on over-simplified multipole moments; in a real nucleus E1 moments are usually two or three orders of magnitude down on this estimate except for the very light nuclei, and E3 and greater moments are down, though not markedly, whereas in the region of deformed nuclei E2 moments can exceed the estimate by a factor ~ 10. (Note that the corresponding rates depend upon the squares of these moments.) Magnetic transitions are down by the factor $(v/c)^2$ in intensity on their corresponding electric transitions and are suppressed rather than enhanced relative to the simple estimate. Thus in practice Coulomb excitation selectively picks out (enhanced) E2 transitions and in odd nuclei it is quite common to observe excitation by the E2 moment followed by M1 γ-radiation when the selection rules allow both moments. Obviously the strength of the interaction contains the product $Z_1 Z_2$, so it pays to go to higher charged projectiles (note that the parameter $\omega a/v$ also increases as $Z_1 Z_2$; but here it is offset by going to higher E_K, until $2a$ is reduced to a value somewhat greater than $(R_1 + R_2)$ which changes rather slowly with the higher A corresponding to the higher charge on the projectile). In fact for projectiles with $Z > 10$ or so and appropriate kinetic energies, the interaction is so strong that, while the projectile is still moving towards the nucleus, the excitation from the ground state is so great that the excited state becomes an effective ground state for further excitation and so on, with the result that, by the time the projectile has passed, the nucleus may be left in any level in the ground state band up to as high as the fifteenth member ($J = 30$ for a $J = 0$ ground state). The process is known as multiple Coulomb excitation and usually occurs by way of enhanced E2 transitions.

Using this technique the rotational bands shown for ^{238}U and ^{177}Lu in Fig. 3.7 have been extended up to higher energies than is shown in that figure; ^{238}U for example has the 24^+ member of the ground state band at 3.5 MeV approximately. At this region of excitation there must be a host of background states since the pairing energy (which is responsible for O^+ ground states of even–even nuclei) is only about 1 MeV in such heavy nuclei, and three broken pairs can recouple in an enormous number of ways. How then is it possible to infer that the above process has taken place and that a state of ~3.5 MeV is the 24^+ member of the ground state band? The reason is that high angular momentum is attained at minimum energy when the nucleus as a whole (or at least a large part of it), with its high moment of inertia, is made to rotate, rather than by sharing this angular momentum among a few nucleons in single particle states. So for $J \geqslant 12$ (to be conservative; the statement is probably correct for all even J and positive parity in even-even nuclei), each member of the ground state band is the lowest lying state for that particular J—such states are known as yrast states. The state 24^+ will not be able to decay by E1 or M1 to states 23^- or 23^+ (or another 24^+) since there are no such states available at lower energy (they would need to be members of rotational bands built upon states 1^- or 1^+ and these lie quite high above the ground state); since the E2 transition to 22^+ is enhanced this will be virtually the only mode of decay. Thus a characteristic of multipole Coulomb excitation is a cascade of γ-radiation down the band making detection and enumeration of the states relatively simple. This argument also indicates that these states will be virtually pure members of the band since there are no other states of the same J^π near enough to mix in with them. The simple purity of yrast levels has attracted the attentions of experimentalists, and has given much information about states at high excitation.

Stripping and Coulomb excitation are typical direct reactions, of which there are many other examples. The inverse (p, d) or pick-up reaction, (^{3}He, d), (α, p or d or T), etc., and their inverses are further examples of stripping. Other direct reactions are elastic and inelastic scattering of charged particles (usually spinless, e.g. α-particles) by deformed nuclei or nuclei capable of vibration, and transfer reactions e.g. (^{17}O, ^{16}O) or (^{16}O, ^{12}C) in which a single nucleon or small cluster of nucleons is transferred from the projectile to the target nucleus or vice versa—though strictly speaking these could be classed as stripping with heavy-ion projectiles.

Problems

6.1. For elastic scattering of a projectile of mass m by a target of mass M show that the angle of scattering in the laboratory system (θ) and the centre-of-mass system (ϕ) are connected by $\phi = \theta + \sin^{-1}\{(m/M)\sin\theta\}$ and that the ratio of solid angles between the two systems is given by $d\Omega_{CM}^{(\phi)}/d\Omega_{lab}^{(\theta)} = \sin^2\phi/\sin^2\theta\cos(\phi-\theta)$. Explain the significance of these transformations in the measurement of differential cross-sections.

6.2*. A foil of $^7_3$Li is bombarded with protons of 5.0 MeV, and it is found that at 90° neutrons of 2.3 MeV are emitted. If the atomic masses of p, n, and ^{7}Li are respectively 1.0078 u, 1.0087 u, and 7.0160 u find the atomic number and mass of the daughter nuclues.

6.3*. A thin target of ^{27}Al is bombarded with 8 MeV protons, and the energy spectrum of the protons scattered at 45° to the incident beam contains strong peaks at 7.84 MeV and 6.99 MeV. What can be deduced about the levels of this nucleus?

6.4. By using an argument involving parity, show that $J = \frac{1}{2}$ for the intermediate state in a γ–γ cascade will, like $J = 0$, result in an isotropic angular correlation. Show also that the γ-decay of an isomeric $J = \frac{1}{2}$ state will be isotropic even when the state has been completely polarized into one of its sub-states (e.g. by application of a magnetic field at low temperature), but will not be isotropic in this latter condition if the γ-detector is sensitive to the state of circular polarization.

6.5*. In a given nucleus, a state C (0^+) decays by γ-cascade through a state B to the ground state A (0^+). If the two γ-transitions are E1, what is the spin and parity of B? Will there be a non-isotropic angular correlation of the two γ-rays? Alternatively, if the γ-radiations are again E1 but B is known to be 0^+, what are the spins and parities of A and C? Will there be an angular correlation?

For the more general case, discuss qualitatively how angular-correlation measurements can help to give information about spins and parities of nuclear states: (a) if the multipole order of each γ-ray is known (e.g. from internal conversion measurements); (b) if it is not.

6.6*. Suppose that the reaction ^{12}C(α, γ) ^{16}O ($Q = 7.15$ MeV) shows an isolated resonance at $E_\alpha = 10.10$ MeV with $\Gamma_{tot} = 0.20$ MeV, $\Gamma_\alpha = 0.15$ MeV and $\Gamma_\gamma = 10$ eV. The spin of the compound state is $J = 4$. The only other decay mode of ^{16}O at this excitation is by proton emission to the ground state ($J = \frac{1}{2}$) of ^{15}N. When the same level is excited by ^{15}N(p, γ) ^{16}O ($Q = 12.11$ MeV) what will the peak cross-section be? (Remember to consider centre-of-mass effects.)

6.7*. (a) Write an equation embodying the principle of detailed balance as it applies to two-body reactions with unpolarized beams and targets. (This should be deduced from Fermi's golden rule and the definition of cross-section—since these have been used to deduce the Breit–Wigner formula it can also be conveniently extracted from the latter, though it has greater generality.) Explain how it may be used to deduce the spin of the pion from the measured cross-sections for $p + p \rightleftharpoons d + \pi^+$. (Note there is in addition a factor arising from the production of two identical particles when the reaction proceeds from right to left.) (b) Explain why the Breit–Wigner formula is important in the theory of slow neutron capture by nuclei. Show that, at a resonance peak (neglecting any background), $\pi\sigma^2 = \lambda^2 g\sigma_s$, where σ and σ_s are the total and elastic scattering cross-sections for neutrons.

7. Fission and fusion

In these days of energy crises no textbook on nuclear physics would be complete without some reference to these two sources of energy. Fusion has very little overlap with nuclear physics, except for the fact that the basic reactions are between light nuclei, but the nuclear reactor based on fission can truly be said to be the brain-child of the nuclear physicist. Indeed virtually the whole of nuclear physics is involved in describing how and why a reactor works.

The fission process

In Chapter 1, the mass formula indicated that all nuclei above $A \sim 150$ are in principle unstable to α-emission. At the same time, the question of stability against splitting up in any way whatsoever might have been considered, but was not. It is so now. Splitting into two approximately equal parts is expected to be possible at large enough mass values, since the mass formula indicates that medium-weight nuclei are most stable. Expressing $M(Z, A) - M(Z_1, A_1) - M(Z_2, A_2)$, where $Z = Z_1 + Z_2$, $A = A_1 + A_2$, in terms of the mass formula (p. 16), and maximizing this function, gives the result $Z_1 = Z_2$, $A_1 = A_2$, and

$$M(Z, A) - 2M(\tfrac{1}{2}Z, \tfrac{1}{2}A) = -\beta A^{\frac{2}{3}}(2^{\frac{1}{3}} - 1) + \varepsilon Z^2 A^{-\frac{1}{3}}(1 - 2^{-\frac{2}{3}}),$$

neglecting the trivial term in δ. The condition for stability is therefore

$$\frac{Z^2}{A} < \frac{\beta}{\varepsilon}\left[\frac{2^{\frac{1}{3}} - 1}{1 - 2^{-\frac{2}{3}}}\right] \sim 0.7\left(\frac{\beta}{\varepsilon}\right) \sim 17.$$

Thus, in principle, all nuclei above $A = 90$ $(Z = 40)$ are unstable to fission into equal fragments. This low value is reminiscent of the low value obtained for α-decay, which in turn leads to the reason why such elements do not fragment, namely, the Coulomb barrier. It might be more realistic to expect fission to occur if the mass of the nucleus under consideration exceeds the masses of the two parts by more than the Coulomb energy when the two parts have just separated. Fission will then occur over the top of its barrier. This adds a term

$$\frac{(Ze/2)^2}{4\pi\varepsilon_0 \cdot 2R_0(A/2)^{\frac{1}{3}}}$$

to the energy of the two fragments. Inserting a suitable value for R_0 (1.3 fm), or alternatively assuming that ε derives from a uniform charge distribution and using the value of ε to determine the radius, modifies the condition to $Z^2/A > 50$ for instability, and this would correspond to an extrapolation to $Z \sim 130$ along the valley of stability.

These estimates are very crude. It has been assumed that, in inverse, as they move toward each other the two fragments remain spherical up to contact. But nuclei in isolation can be non-spherical, and even if spherical it is not realistic to assume they are rigid. In the presence of each other it is possible that the Coulomb potential be less than the monopole term predicts because of induced distortion or because the proton distribution inside each nucleus may no longer be uniform. These and other mechanisms can diminish the Coulomb potential near the point of contact whilst not greatly affecting the potential of the single large nucleus. It is to be expected that spontaneous fission will occur for less massive nuclei than predicted by the above formula.

To consider the mechanism in more detail, a potential function is needed to describe the interaction between the two fragments. When the two components are distinct, the potential can be expressed as a function of the distance between their centres, but this parameter has little meaning before separation. In this region a convenient parameter is ΔR for an ellipsoidal nucleus as defined in Chapter 1 (p. 26), but this ceases to be useful near the point of break-up since $\Delta R \to \infty$ produces a needle-like nucleus rather than two globules. If we ignore this difficulty the potential is shown in Fig. 7.1, where the two dotted curves correspond to

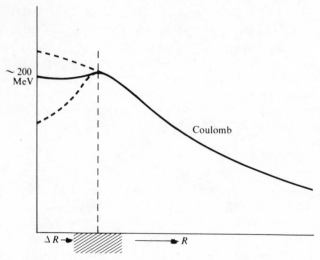

Fig. 7.1. Schematic fission potentials. Note that the abscissa is the distortion parameter ΔR up to somewhere below separation and the distance between fragment centres above separation. The dashed curves are explained in the text.

the limiting cases of a nucleus in principle unstable to fission not observed to decay so and of a nucleus which would decay by the fission process in a time of order of the nuclear characteristic time. The interesting case is shown by the full line which represents a typical distorted nucleus in the region of uranium.

If the well representing the large nucleus is extremely shallow then it is possible for states inside it to decay by penetrating the fission barrier (since the particle penetrating has large effective mass, the penetrability varies extremely rapidly with energy below the top of the barrier so that shallow means of the order of a few hundred keV). This leads to a class of radioactivity not previously discussed, namely, decay by spontaneous fission. If the well is deeper than this, then the levels near ground may be stable to this mode of decay, but levels of higher excitation will not be, so fission will take place in preference to γ-emission. Such levels could be excited by inelastic scattering, e.g. (n, n') or (p, p'). At a greater well-depth, the location of the top of the fission barrier relative to neutron instability is of great importance. If the fission barrier lies lower then the nucleus one neutron less than the one here considered, on absorbing a thermal neutron, will decay by fission; if the fission barrier lies say 1 MeV higher then neutrons of energy greater than 1 MeV will be necessary to induce fission. Notice that a neutron threshold is well defined, but the fission threshold is effective in that just below it the process can take place but with very much reduced probability.

As an example of the above, spontaneous fission occurs in the following:

$$^{235}_{92}U \ (\tau_f \sim 2 \times 10^{17} \ \text{years}), \qquad ^{238}_{92}U \ (\tau_f \sim 10^{16} \ \text{years}),$$

$$^{240}_{94}Pu \ (\tau_f \sim 10^{11} \ \text{years}), \qquad ^{250}_{98}Cf \ (\tau_f \sim 10^{4} \ \text{years}),$$

$$^{254}_{98}Cf \ (\tau_f \sim 55 \ \text{days}).$$

Nuclei which undergo fission on capture of thermal neutrons are ^{233}U, ^{235}U, and ^{239}Pu, whilst nuclei which require capture of fast neutrons are, with effective neutron thresholds in brackets, $^{232}_{90}Th$ (1.3 MeV), ^{234}U (0.4 MeV), ^{236}U (0.8 MeV), ^{238}U (1.1 MeV), and $^{237}_{93}Np$ (0.4 MeV). The two last sequences reflect the importance of even or odd neutron numbers, emphasizing the effect of pairing of nucleons.

For nuclei in this region the fission reaction is highly exothermic with a Q-value of order 200 MeV. This is highly desirable for the production of energy, but of itself does not lead to power production. The reaction must be self-sustaining as in chemical combustion when part of the energy liberated in the reaction is used to heat the fuel and so encourage further reaction. From information now to be disclosed, the fission reaction can be made self-sustaining. Fission is not merely the splitting of a large nucleus into two smaller ones but, as is to be expected, a lot of debris is produced at the same time. Neutrons form part of this debris, either produced simultaneously with the large fragments or more probably by secondary emission from the fragments which will in general be formed in states of high excitation. In addition, the fragments lie on the neutron-rich side of the valley of stability (since N/Z increases with Z along the valley) and therefore have low binding energies for neutrons. These neutrons will be emitted almost instantaneously ($\tau \sim 10^{-16}$ s or less) with a spectrum peaked at ~ 1 MeV. This latter point can be understood in terms of the discussion on widths in Chapter 6; Γ_n from a given state to any final nuclear state is proportional to k or $E_n^{\frac{1}{2}}$, but the density of final nuclear states has very roughly an exponential dependence upon energy of excitation. The most probable energy of emission of a neutron is therefore the maximum of $(E - E^*)^{\frac{1}{2}} \exp(\alpha E^*)$, where E is the energy of the neutron transition to ground and E^* the excitation of the final nucleus. At maximum $E - E^* = 1/2\alpha$. Peaking at $E_n = E - E^* \sim 0.7 \, \text{MeV}$ gives $\alpha = 0.7$ in units of $(\text{MeV})^{-1}$, indicating how the level density behaves for a heavy nucleus. Note that $\exp(\alpha E^*)$ is an oversimplification, but it does represent roughly the crowding together of levels at higher energies; α varies slowly from nucleus to nucleus.

In addition to these 'prompt' neutrons there are also emitted some delayed neutrons amounting to ~ 0.5 per cent of all neutrons produced. These were initially a source of mystery, but are now known to occur following β-emission; an example is the decay of the fission fragment $^{87}_{35}\text{Br}$ which β-decays with 55 s half-life to $^{87}_{36}\text{Kr}$, with an available energy $\sim 6 \, \text{MeV}$. As well as decaying to nucleon stable levels in ^{87}Kr, a fraction of the β-transitions are to states of excitation greater than $5.4 \, \text{MeV}$, and these can decay by neutron emission. Neutron emission will therefore be delayed by the time taken for β-decay of ^{87}Br. Although relatively few, these delayed neutrons play a large part in the control of a reactor. If each neutron-induced fission event produces on average more than one neutron, the possibility of a self-sustaining chain reaction arises. Typically the number ν, per fission (average) is something like 2–2.5.

Before going on to discuss how to make use of the fission reaction, comment should be made on the reference to the fragment $A = 87$. This is a long way from the symmetric splitting $(A \sim 115)$ which has been suggested as giving the greatest energy release. In fact the most probable mode of fission is not the symmetric mode, but corresponds roughly to an $A = 90$ and 140 split. The explanation of this is still required; magic numbers probably play a part since $N = 50$ for $A \sim 90$, $N = 82$ for $A \sim 140$. Note that the Coulomb term is less for the asymmetric split relative to the symmetric, offsetting to some extent the mass balance which favours the symmetric split. Prediction of the 'mass spectrum' requires detailed knowledge of the fission process—any mechanism which can change the fission barrier height, as a function of the fragment masses, by $\sim 0.1 \, \text{MeV}$ in

200 MeV could account for the spectrum. It is interesting to note that, close to threshold, the (90, 140) splitting predominates and there is virtually no symmetric splitting. As the energy of the neutron inducing fission rises, the splitting becomes more nearly symmetric in accordance with the rough guidelines above.

The chain reaction and reactors

If a material fissile to fast neutrons produces v neutrons per fission (on average) and the probability that, on absorbing a neutron (and the definition can be extended to include absorption in other materials present), fission occurs is p (again on average) then, if $pv > 1$, a chain reaction will occur in a large mass of this material, culminating in an explosion. The intensity of the explosion depends on the rate of liberation of energy compared with the rate at which energy can be transmitted away (possibly taking the material with it). If the interaction length for fission is λ then the interaction time is $\lambda/\bar{v}$, and in that (mean) time the neutron multiplication is $(pv - 1)$, giving an exponential growth in neutron flux with time constant $\lambda/\bar{v}(pv - 1)$. For ^{235}U, $\sigma_f \sim 1\,\text{b}$, giving an interaction length $\sim 20\,\text{cm}$, $\bar{v} \sim 10^9\,\text{cm s}^{-1}$ $(E_n \sim 1\,\text{MeV})$, and for $pv - 1 \sim 1$ this gives a time constant $\sim 10^{-8}\,\text{s}$. For such a rapid rate of neutron growth and therefore energy production, heat dissipation is negligible, and because of inertia the material has no time to get away. The result is the so-called 'atom bomb'. From what has been said the critical size for explosion is something greater than 20 cm, and this can be achieved by imploding (chemically) some smaller blocks of ^{235}U.

On the face of it, it seems unlikely that such a system could be controlled. The fissile material could be diluted and pv could be made very close to unity, but it would be difficult to produce just the correct conditions for a slow enough growth rate in order to be able to introduce some form of mechanical control, and if by error conditions are changed to make pv slightly greater then the system runs amok. At this point the fact that ~ 0.5 per cent of the neutrons are delayed becomes crucial; if pv is just less than unity for prompt neutrons and just greater than unity for prompt-plus-delayed neutrons then the time constant of the delayed neutrons takes control and the growth-rate time constant will be considerably greater than this (since $pv - 1 \sim 0.005$). Thus there is a range of variation of pv of ~ 0.5 per cent over which the neutron flux is increasing at a controllable rate. For the fast reactor, which is effectively being discussed here, control would be achieved by moving parts of the fissile material away from the rest or by moving some of the surrounding material relative to the fissile core, thereby altering the number of neutrons being reflected back into the core.

Historically the thermal reactor preceded the fast reactor. Why was this the case? The answer lies in understanding what makes a suitable fuel. Although ^{238}U and some other nuclei listed on p. 139 are fissionable above a neutron threshold they are not suitable fuels for a fast reactor, whereas nuclei like ^{235}U, which are fissionable to thermal neutrons, are. For these latter nuclei the factor p depends only upon the ratio $\sigma(n, f)/\sigma(n, \gamma)$, since $\sigma(n, n')$, which is larger than $\sigma(n, \gamma)$ at MeV energies, does not remove neutrons but merely changes their energies. However, if ^{238}U were used as a fuel then (n, n') would rapidly remove neutrons below the fission threshold and so be akin to absorption, with a reduction in p. This process is enhanced by the fact that the reactor has to have a structure and a coolant so the neutron spectrum inside a fast reactor is degraded to lower energies very quickly. It is therefore necessary that the fuel be fissile down to low energies. Thus natural uranium, containing only 0.7 per cent ^{235}U, cannot be made to go critical; the fuel must be highly enriched in ^{235}U, a costly undertaking.

It would therefore seem surprising that any structure containing natural uranium can be made to go critical. If, however, we look at the various cross-sections at thermal energies (~ 0.025 eV) we find that, for ^{235}U, $\sigma(n, \gamma) \sim 101$ b and $\sigma(n, f) \sim 577$ b, whereas for ^{238}U, $\sigma(n, \gamma) \sim 2.76$ b and $\sigma(n, f) \sim 0$. Even taking into account that there is 140 times more ^{238}U than ^{235}U in natural uranium, a value $p \sim 0.54$ is obtained for thermal neutrons. Since $v \sim 2.44$ for ^{235}U, it is possible to obtain criticality provided that not more than a quarter of the neutrons are lost in thermalizing them. Before considering the thermalization process it is pertinent to ask why $\sigma(n, \gamma)$ for ^{238}U turns out to be so much smaller at thermal energies than the ^{235}U cross-sections. This is partly a fortunate accident, but not completely so.

On comparing the excitation functions of $U + n$ for the

different isotopes, an odd–even effect is immediately apparent. The even isotopes produce narrow, well-separated resonances having very high peak cross-sections; the odd isotopes produce broader resonances, poorly separated (the density of resonances is ~10 times greater for the odd than the even isotopes) and with lower peak cross-sections. Previously it has been stated that level density increases exponentially with excitation, and the odd–even effect does ensure that the excitations produced by absorption in an odd isotope are about 1.5 MeV higher than in an even isotope reflecting the fact that in the former case the resulting ground state has one more pair of neutrons than the initial ground state (^{235}U + n_{th} gives 6.48 MeV excitation, ^{238}U + n_{th} gives 4.78 MeV). But, because pairs must be split in order to create excited states there is an odd–even effect needed in applying the exponential function. It is probably more correct to translate energies such that neutron thresholds coincide in comparing level densities of neighbouring isotopes. There is, however, a $(2J + 1)$ factor appearing in the level-density function, and this can account for the observed ratio, since ^{235}U + n produces states of $J^{\pi} = 3^-$ and 4^- (^{235}U ground state is $\frac{7}{2}^-$ and $l_n = 0$ for neutrons in the eV range) whilst ^{238}U + n produces $J^{\pi} = \frac{1}{2}^+$ only. Merely increasing the level-density ratio should not (on average) enhance the capture cross-section ratio at thermal energies, since $\sigma_{n\gamma} \propto \Gamma_n\Gamma_\gamma/E_r^2$ (assuming that $E_r > \Gamma$— see the Breit–Wigner expression, p. 123), and this can be written

$$\frac{(\Gamma_n/D)(\Gamma_\gamma/D)}{(E_r/D)^2}$$

Provided the top two factors remain much the same from one isotope to the other it would appear that the thermal cross-section depends upon how close is the nearest level compared to the level spacing—and this is how chance creeps in. However, though (Γ_n/D) is not expected to vary much between neighbouring nuclei since it refers to a single channel; the same is not true for (Γ_γ/D), because it represents a summation over a large number of γ-transitions. The same $(2J + 1)$ factor, which produced a greater density of resonances in ^{235}U + n, also provides a greater density of states to which γ-transitions can be made since strong γ-transitions change J only by one unit at most. It therefore turns out that total radiation widths Γ_γ, following neutron absorption, tend to be similar for neighbouring nuclei, rather than Γ_γ/D. There is therefore a tendency for even-A isotopes to have lower thermal-capture cross-sections than odd-A isotopes, apart from any odd–even effects which may also occur because of pairing. This argument is little affected by the presence or absence of fission width, though it may need slight modification if the increased total width no longer satisfies $E_r > \Gamma$.

Returning to thermalization, neutrons always lose energy in elastic collisions, and the lighter the scattering nucleus the greater the loss. Hydrogen would therefore seem to be the most suitable material, conveniently perhaps in the form of water; but, unfortunately it has a relatively high thermal-neutron capture cross-section (~ 330 mb), and so absorbs many of the neutrons on slowing them down. It cannot be used with natural uranium, but uranium enriched in ^{235}U can go critical in a light (ordinary) water environment, and modern power reactors have been constructed to this basic design. The next suitable material is heavy-water since ^{2}H is the next best thermalizer, and, moreover, $\sigma(n, \gamma)$ is very low (0.5 mb). Again, modern power reactors have been built using natural uranium fuel and heavy water. The next most convenient material is carbon, which is much poorer at moderating but still has a small enough $\sigma(n, \gamma)$ to be of value (3.4 mb). The early experimental reactors were built with carbon (graphite) as moderator and natural fuel, and modern power reactors have been so based.

The geometrical arrangement of fuel and moderator is important. The obvious homogeneous mixture is unsuitable for natural fuel since the neutron capture process in ^{238}U is important right down to its lowest resonance at $E_n \sim 6.6$ eV (see Fig. 6.5, p. 125). Separation of the two processes of fission and moderation is needed, but the next obvious step of having all the fuel at the centre surrounded by moderator is also unsuitable. At thermal energies the effective cross-section for removal of neutrons is $2.7 + (101 + 577)/140 \approx 7.5$ b, weighting the cross-sections according to the isotopic abundance, giving an absorption length of order 2 cm. Thus the thermal-neutron flux into the fissile centre is strongly attenuated, and most of the fuel is inefficiently used. The solution is therefore a matrix of natural fuel rods a few millimetres in diameter embedded in the moderator, through which must also be circulated a coolant with low neutron-absorbing properties to transfer the heat produced to the outside world. At least this is the case for the graphite-moderated reactors, the suitable coolant being carbon dioxide gas; the liquid moderators can, in addition, be used for heat transfer. Note that, because the fission fragments have such a short range, virtually all the heat is produced within the fuel itself, with a small fraction arising from slowing down the neutrons and absorption of γ-radiation.

The control of a thermal reactor derives once again from the existence of delayed neutrons, though the time constant when critical to prompt neutrons is rather slower because the neutrons, when thermalized diffuse rather slowly through the moderator (thermal speeds are $\sim 10^5 \, \mathrm{cm \, s^{-1}}$ so diffusion speeds are lower and distances of traverse to get back to fuel are $\sim 10 \, \mathrm{cm}$ giving a (prompt critical) rise time $\sim$ ms). The mechanical control is in the form of cadmium rods that can be inserted into the reactor. Cadmium is a very strong absorber of thermal neutrons by virtue of the large value of $\sigma(\mathrm{n}, \gamma)$ ($\sim 20,000 \, \mathrm{b}$) for the isotope $^{113}_{48}\mathrm{Cd}$, of abundance 12.3 per cent. This high cross-section arises from a resonance very close to thermal energies, as has been discussed previously for $^{235}\mathrm{U}$.

So far only the rare isotope $^{235}\mathrm{U}$ has been utilized as a nuclear fuel. Obviously if $^{238}\mathrm{U}$ could also be used resources of power would be considerably increased. This can be done, making use of the very process that has proved troublesome in the design of reactors discussed, namely, the capture process to form $^{239}\mathrm{U}$. $^{239}\mathrm{U}$ undergoes two quick β^--decays to $^{239}\mathrm{Pu}$, which is long-lived ($\sim 24,000$ years) and has already been listed among nuclei which are thermally fissile. This breeding of nuclear fuel can be best achieved with a fast reactor since ν tends to rise with neutron energy and the greater its value the greater the surplus of neutrons for breeding over and above the requirements for sustaining the reaction. This surplus is used by making the core small enough to allow a large leakage of neutrons into a surrounding uranium blanket. Since the values of $p\nu$ are 2.45 and 2.70 for pure $^{235}\mathrm{U}$ and $^{239}\mathrm{Pu}$, even after reasonable loss incurred from other materials and coolant, within the core, it is possible to breed more new fuel than is consumed. It is necessary, of course, to remove the uranium blanket from time to time to separate out chemically the ^{239}Pu, which is behaving effectively as a catalyst in the fission of ^{238}U. Thus, in principle, all of natural uranium can be usefully consumed and at ~200 MeV per nucleus this works out at about 1 g per day per MW!

Fusion in the Sun

Before discussing the nuclear physics behind attempts to produce fusion in the laboratory, it is of interest to describe briefly the nuclear reactions going on inside the nearest working fusion reactor, the Sun. Prior to the discovery of nuclear reactions the Sun's source of energy was a mystery. The rate of emission of radiant energy from the Sun is well known and so also is its mass; gravitational collapse of this mass provides a large source of energy, but nowhere near large enough to keep the Sun radiating for a few thousand million years, and there is geological evidence on Earth to indicate that it has done so for a considerable fraction of that time. On the other hand, gravitational collapse to the present size from a tenuous gas will provide enough energy to heat the Sun up to a temperature $\sim 10^7$ K even allowing for radiation losses in the slow process. At that temperature, corresponding to a mean thermal energy of ~1 keV, protons are beginning to collide sufficiently violently with each other to penetrate the Coulomb barrier. The stumbling block to the onset of nuclear reactions culminating in the synthesis of helium from hydrogen would appear to be the initial reaction of two colliding protons. If we assume that two protons momentarily make a compound nucleus, the diproton, then this nucleus will decay back into the entrance channel thereby producing no reaction. However, a parallel though improbable decay is by β^+-emission to the deuteron. Improbable though it may be, the Sun contains a lot of colliding protons which cannot disappear by any other process, so this one must occur (though it has never been observed in the laboratory) initiating the sequence:

$$p + p \rightarrow d + e^+ + \nu$$

$$p + d \rightarrow {}^3\text{He} + \gamma$$

$${}^3\text{He} + {}^3\text{He} \rightarrow {}^4\text{He} + 2p$$

Taking the first two reactions twice, the nett result is

$$4p \rightarrow {}^4He + 2e^+ + 2\nu + Q, \tag{7.1}$$

and from the mass values this liberates ~7 MeV per proton absorbed (by comparison fission gives less than 1 MeV per nucleon involved). The choice of reactions after the first is dictated by the preponderance of protons. Thus the possible third reaction $d + {}^3He \rightarrow {}^4He + p$ will not occur since all deuterons formed will be mopped up by the second reaction before they can get near a ${}^3He$ nucleus. On the other hand, the ${}^3He + p$ interaction can only decay back into the same channel, so ${}^3He$ is stable in the presence of protons and will build up in concentration until the third reaction takes place at the appropriate rate for equilibrium.

The rate at which a reaction proceeds depends upon $n_1(v_1)n_2(v_2)\sigma(v_{12})v_{12}$, where n_1 and n_2 are velocity distributions of densities of particles 1 and 2 per unit energy range and v_{12} is their relative velocity. The two Maxwellian distributions can be combined into a single such function in terms of the reduced mass and relative velocity giving (dropping the suffixes) $n_1(v_1)n_2(v_2)v_{12} \propto v^3 \exp(-mv^2/2kT)$. $\sigma(v)$ can be written as $\sigma_0 \cdot P(E) \cdot (E_B/E)^{\frac{1}{2}}$, where $P(E)$ is the Gamow factor, $\exp(-2\pi Z_1 Z_2 e^2/4\pi\varepsilon_0 \hbar v)$, in the penetrability for the limiting case of an energy E well below the barrier height E_B, and therefore for a nuclear radius which is very small compared to the classical distance of closest approach. $(E_B/E)^{\frac{1}{2}}$ includes the $\pi \lambda^2$ term in the cross-section. Expressed in this way σ_0 is the cross-section at $E \sim E_B$ and is of the order of 0.1 b for a decay by particle emission and perhaps ~10 μb for decay by γ-emission.

The reaction rate is obtained by integrating over all v. The two key factors will be the rapidly falling $\exp(-E/E_T)$, where E_T, the thermal energy, is ~1 keV at 10^7 K, and the rapidly rising $\exp(-\beta E^{-\frac{1}{2}})$ where, for $Z_1 = Z_2 = 1$ and E in keV, $\beta \sim 30$. Maximizing the product of these two functions (the rate of variation of other factors can be neglected) gives $E^{\frac{3}{2}} \sim \beta E_T/2$, giving $E \sim 6$ keV. Thus in the region around 6 keV there will be a sharp peak in the reaction rate (see Fig. 7.2). At this energy the penetrability factor is $\sim \exp(-12)$ or 10^{-5} and the cross-section for $p + d \rightarrow {}^3He + \gamma$ will be $\sim 10^{-10}$ b; the reaction ${}^3He + {}^3He \rightarrow {}^4He + 2p$, having $Z_1 = Z_2 = 2$, will peak at ~15 keV

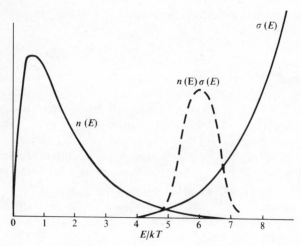

Fig. 7.2. An indication of the peaking of the yield curve of a thermonuclear reaction.

and give a cross-section $\sim 10^{-10}$ b. For the $p + p$ reaction the cross-section will in addition contain a factor representing the probability of β^+-decay during a nuclear characteristic time. A β^+-lifetime $\sim 10^3$ s is a rough guess (about the same as the β^--decay of the neutron), and a nuclear time $\sim 10^{-21}$ s gives the factor $\sim 10^{-24}$ and a cross-section at peak of about 10^{-29} b. Notice that none of these cross-sections is measurable in the laboratory; measurements could be pushed down to ~ 30 keV in the second and ~ 100 keV in the third reaction and the energy dependence used to extrapolate further down. In the first reaction the factor 10^{-24} precludes measurement at any energy so the cross-section must be derived theoretically.

An alternative set of reactions to the above is the carbon–nitrogen cycle:

$$^{12}\text{C} + \text{p} \rightarrow {}^{13}\text{N} + \gamma; \qquad {}^{13}\text{N} \rightarrow {}^{13}\text{C} + e^+ + \nu$$
$$^{13}\text{C} + \text{p} \rightarrow {}^{14}\text{N} + \gamma; \qquad {}^{14}\text{N} + \text{p} \rightarrow {}^{15}\text{O} + \gamma \qquad (7.2)$$
$$^{15}\text{O} \rightarrow {}^{15}\text{N} + e^+ + \nu; \qquad {}^{15}\text{N} + \text{p} \rightarrow {}^{12}\text{C} + {}^4\text{He}$$

Thus the ^{12}C is regenerated and acts as a catalyst. The importance of the two cycles depends upon the temperature and upon the amount of ^{12}C present. (Presumably the amount present in the primordial gas from which the Sun condensed, since formation of heavier elements requires a considerable concentration of ^{4}He in order to start the process with $3\,^4$He $\to\,^{12}$C $+ \gamma$. This reaction is necessary since p $+\,^4$He merely returns to the same channel, as does ^{4}He $+$ He^4.) At high enough temperatures (more massive stars) this latter cycle will take over, since none of the processes is particularly inhibited except for the Gamow factor, which will be larger because of the increased charges on C and N.

Fusion in the laboratory

Neither of the cycles described are suitable bases for terrestrially controlled fusion. The hydrogen cycle begins with the improbable (weak) reaction of two protons whilst the carbon–nitrogen cycle requires higher temperatures which must be maintained over periods long compared with the 10 min lifetime for β^+-decay of ^{13}N. The chief problems are the heating up of the reacting materials, and, more difficult, the containment of the hot material whilst sufficient reaction occurs to make a nett liberation of energy. A gas at 10^7 K (or strictly a plasma, since it will be highly ionized) cannot simply be retained by walls since, no matter what the material of the container, it will not be able to withstand bombardment of particles at such energies either physically or chemically. Much work has been carried out using 'magnetic bottles'; since the particles in the plasma are charged they are deflected by magnetic fields, and a great deal of cunning has gone into designing the field so that the particles are kept away from the walls of the vessel. In fact the magnetic fields can be used to feed energy into the plasma and to compress it into a smaller and smaller volume thereby increasing the reaction rate both by temperature and density increase. Unfortunately, the system becomes unstable, and so containment cannot be maintained indefinitely. Containment times are at present measured in fractions of a second, which of course can be large enough to liberate enormous amounts of energy as in the atom bomb or its later development the hydrogen bomb, which is merely an atom bomb surrounded by fusible material; the atom bomb raises the temperature high enough ($\sim 10^9$ K) and fusion then occurs more rapidly than the energy and the material can be dissipated.

Obviously the reaction, or cycle of reactions, must be

intrinsically fast. A suitable one for the laboratory would be

$$d + d \rightarrow {}^3He + n$$
$$\rightarrow T + p, \qquad (7.3)$$

since it uses deuterium only which occurs as 1/7000 of natural hydrogen and so is available in unlimited supply. The reaction involves unit charges and will therefore go at the lowest necessary temperatures (10^7 K in the Sun, but probably 10^8 K for short containment times). These two reactions have rather low Q-values (+3.27 MeV and +4.03 MeV), releasing only ~1 MeV per nucleon. In addition, they have relatively low peak cross-sections, ~100 mb peaking at several hundred keV. On the other hand, the secondary reactions which could occur, namely $d + {}^3He \rightarrow p + {}^4He$ and $d + T \rightarrow n + {}^4He$, have higher Q-values (18.4 MeV and 17.58 MeV), and higher peak cross-sections, especially the latter which peaks at ~80 keV with $\sigma \sim 7$ b (see Fig. 7.3). This latter reaction therefore, would, go very quickly at 10^8 K, corresponding to a mean thermal energy of ~10 keV, at which energy the cross-section is still as high as 10 mb. This mopping up, if complete, would boost the energy output to ~4 MeV per nucleon.

From the above, the most convenient reaction to solve the problem of obtaining a useful source by fusion is undoubtedly $d + T$ but it may be difficult to use it as a basis for large-scale energy production. In principle, the tritium could be regenerated by the reaction

$$n + d \rightarrow T + \gamma (6.26 \, MeV)$$

or more conveniently by

$$n + {}^6Li \rightarrow T + {}^4He + 4.78 \, MeV.$$

This latter reaction has a high thermal cross-section, so that ${}^6Li$ in the form of a solid compound could form a suitable regenerating blanket, as in some fusion bombs. There is no possibility, with these reactions, of breeding more tritium than is consumed but, by addition of beryllium into the blanket, neutron multiplication can result from the dominant reaction

$$^9Be + n \rightarrow \alpha + \alpha + 2n - 1.66 \, MeV,$$

leading to the possibility of tritium breeding.

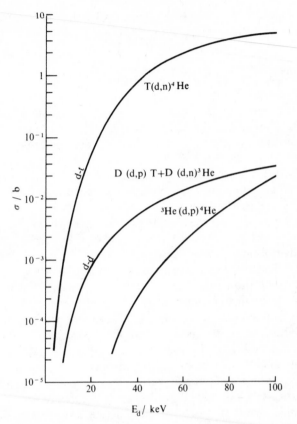

Fig. 7.3.   Thermonuclear reaction cross-sections. (Based on Segre (1964). *Nuclei and particles,* Benjamin, New York.)

A great deal depends upon the supply of materials (assuming that the technical problems are solved). The abundances in the earth's crust in parts per million are, for some of the relevant elements:

$$H, 1400(^2H, 0.2); Li, 65(^6Li, 4.8); Be, 6;$$
$$\text{and compare with U, 4 } (^{235}U, 0.03).$$

These figures do not give a complete picture, since hydrogen is readily available from water (the sea) and deuterium separation is probably the simplest of all isotope separations. Unless the breeding of tritium is relatively simple, it would seem that tritium-induced fusion will only be a stepping stone towards the final solution based on the d + d reaction.

Problems

7.1. If neutrons (mass m) of kinetic energy E_0 are isotropically scattered, in the centre-of-mass system, by nuclei of mass M show that, in the laboratory system, the scattered neutron energy spectrum is independent of energy within the limits E_0 and $a^2 E_0$ where $a = (M - m)/(M + m)$. Show further that the mean value of $\ln(E_0/E_1)$, where E_1 is the energy after a single scatter, is equal to $1 + (a \ln a)/(1 - a)$, denoted by ξ.

7.2. Extend the above result to the case where the initial neutrons are not well defined in energy but have a spectral distribution. Hence show that, after n collisions, $\overline{\ln(E_0/E_n)} = n\xi$. If the mean energy of the neutron fission spectrum is 2 MeV, show that ~18, 25, and 117 collisions are required to thermalize neutrons in H, D, and ^{12}C respectively.

7.3. An intimate mixture of ^{235}U and graphite is required for certain experiments. The graphite is known to be contaminated with 1 p.p.m. by weight of ^{10}B. What is the maximum fraction by weight of ^{235}U in the mixture if the multiplication factor at infinite size is not to exceed unity? ($\sigma_{abs} = 0.04$ b, 3800 b, and 700 b respectively for ^{12}C, ^{10}B, and ^{235}U at thermal energies; of the last cross-section, σ_f accounts for 580 b. Assume 2.5 neutrons per fission and that all reactions take place at thermal energies only.)

7.4. For a gas consisting of two types of particles, show that the distribution of *relative* velocities of pairs consisting of one from each type is the same as the absolute velocity distribution for a single type of particle but with a mass equal to the reduced mass, $m_1 m_2/(m_1 + m_2)$, where m_1, m_2 are the masses of the two types.

7.5. (a) If the fusion reaction d + d → ^{3}He + n + 3.2 MeV takes place with deuterons at rest, what is the kinetic energy of the neutron? (b) If the deuterons must come within 10^{-11} cm of each other, what energy must be supplied to overcome the electrostatic repulsion? (c) If this energy is supplied by heating the deuterium to high temperature, what order of magnitude of temperature is required? (d) Discuss the effect of this heating on the energies of the neutrons produced.

7.6. A nucleus consists of a large number N of particles of spin $\frac{1}{2}$. Determine the number of ways, $R(M_s)$, a given M_s can be produced assuming that $M_s \ll \frac{1}{2}N$ and that the Pauli Principle can be ignored. By identifying the difference $R(X) - R(X + 1)$ with the number of states of total spin $S = X$, show that this number is proportional to $(2X + 1)$. Hence the plausibility of the statement made on p. 143.

Appendix A: Cross-sections

To introduce the concept of cross-section, it is convenient to consider classically a nuclear reaction in which a beam of particles falls on nuclei which behave like completely absorbing spheres of cross-sectional area σ. Assume that one such nucleus is confined somewhere within an aperture of area A through which passes a uniform beam of n particles per second. The number absorbed per second will obviously be $n\sigma/A$. If other nuclei are successively added to build up a practical target of thickness t and nuclear number density ρ (i.e. the number of nuclei per unit volume of target material), then $\rho A t$ nuclei will be exposed to the beam, and the number of incident particles absorbed per second will be $\rho A t n\sigma/A = \sigma n \rho t$, provided that $\sigma \rho t \ll 1$. (If this condition does not hold then the beam flux is decreasing through the target and all nuclei are not exposed to the same beam—under these circumstances the beam decreases exponentially through the target, in the form $n(t) = n(0)\exp(-\sigma\rho t)$.) Notice that the result does not depend upon the aperture area A since, no matter what this area is, provided that the target area is larger, each particle in the beam passes through the same target thickness.

Thus, for the absorption process the number per second absorbed is proportional to the cross-sectional area of the nucleus. Replacing this completely absorbing nucleus by a hard sphere of the same size which scatters all projectiles hitting it, the total number per second scattered out of the beam will be the number per second previously absorbed, $\sigma n \rho t$. The number per second scattered into solid angle $d\Omega$ at (θ, ϕ), in the usual nomenclature, relative to the beam will be $\sigma n \rho t f(\theta, \phi)\, d\Omega$, where $\int f(\theta, \phi)\, d\Omega = 1$, to ensure that the total scattering has the correct value. $f(\theta, \phi)$ is known as the angular distribution of the nuclear process. It is convenient to combine the product $\sigma f(\theta, \phi)$ into a single function $d\sigma(\theta, \phi)/d\Omega$, which is known as the differential cross-section for scattering along the direction (θ, ϕ). Most nuclear reactions are observed under conditions which are characterized by an axis of symmetry in the direction of the beam—so, in general, $\sigma(\theta, \phi)$ becomes $\sigma(\theta)$. Experiments may be carried out under conditions in which two axes are defined, e.g. the beam axis and an axis of polarization, and for these the generalized direction (θ, ϕ) is required.

These simple pictures can be extended to cover more complex situations: the nucleus may not be completely absorbing or scattering but partially the one, partially the other, and partially transparent; and, following absorption, something is usually emitted—the same, or another, type of particle or γ-radiation. These processes can all be included in a general definition of cross-section provided the equivalence of cross-section and nuclear cross-sectional area is revoked. This equivalence has arisen merely from the simple picture used to introduce the concept, and will in any case require modification when the wave nature of the projectile is taken into account. The fact remains, however, that the intrinsic nuclear property which determines the yield of a reaction has the dimensions of an area.

The general definition of cross-section for the reaction $A + a \rightarrow B + b$ is contained in the equation

$$\frac{\mathrm{d}Y_{a,b}(\theta)}{\mathrm{d}\Omega} \, \mathrm{d}\Omega = n\rho t \frac{\mathrm{d}\sigma_{a,b}(\theta)}{\mathrm{d}\Omega} \, \mathrm{d}\Omega, \qquad (A.1)$$

where $\mathrm{d}Y_{a,b}(\theta)/\mathrm{d}\Omega$ is the number per second of particles of type b emitted into unit solid angle at an angle θ to the beam, and $\mathrm{d}\sigma_{a,b}(\theta)/\mathrm{d}\Omega$ is the differential cross-section for the process (a, b).

Also of use is the partial cross-section for the process (a, b) defined as

$$\sigma_{a,b} = \int \frac{\mathrm{d}\sigma_{a,b}(\theta)}{\mathrm{d}\Omega} \, \mathrm{d}\Omega.$$

The total cross-section is defined as $\sigma_{a,\mathrm{tot}} = \sum_b \sigma_{a,b}$, where the summation is carried out over all possible reaction products $B + b$ including the incident channel (known as elastic scattering). This latter cross-section is very useful when the projectiles are neutrons, but not so for charged particles, when it is dominated by the comparatively uninteresting but very large Coulomb cross-section.

Finally, notice that cross-section has been defined in terms of the flux of incoming particles that is their density times their velocity. In calculating reaction rates using perturbation theory the incoming particle wave function is normalized to represent one particle per unit volume, so in going to the cross-section a factor v is needed; σv is a measure of the reaction rate as calculated by perturbation theory. If the incoming particle is just above threshold, it may be possible that the appropriate matrix element varies only slowly with energy, as will also the density of final states if the reaction has positive Q-value. Under these circumstances $\sigma v \sim$ constant or $\sigma \propto 1/v$. This has been deduced in the text from the compound-nucleus formula, but as is seen above it has greater generality.

Appendix B: The need for a pairing term in the mass formula

In order to illustrate the need for a pairing term in the mass formula, it is convenient to neglect all terms in the mass equation but the Coulomb and symmetry terms to simplify these to the form

$$a(A - 2Z)^2 + bZ(Z - 1) \quad (a, b \text{ both positive}).$$

The conditions that (A, Z) is stable relative to its neighbours $(A, Z - 1)$ and $(A, Z + 1)$ are

$$a(A - 2Z)^2 + bZ(Z - 1) < a(A - 2Z - 2)^2 + b(Z + 1)Z$$

and

$$< a(A - 2Z + 2)^2 + b(Z - 1)(Z - 2).$$

These can also be expressed as

$$b \cdot 2Z > 4a\{A - 2Z - 1\}$$

and

$$b \cdot 2Z < 4a\left\{(A - 2Z - 1) + \left(1 + \frac{N}{Z - 1}\right)\right\}.$$

Adding a nucleon to (A, Z) gives:

adding p, the energy change $E_p = -a(2A - 4Z - 1) + b \cdot 2Z$,

adding n, the energy change $E_n = +a(2A - 4Z + 1)$.

The condition for $(A + 1, Z)$ to be stable relative to $(A + 1, Z + 1)$ is that $E_p > E_n$, which simplifies to

$$b \cdot 2Z > 4a\{(A - 2Z - 1) + 1\}.$$

For the second neutron to produce a stable nucleus, the condition is that

$$M(A + 2, Z) < M(A + 2, Z + 1),$$

which is equivalent to increasing A to $A + 1$ in the last inequality, i.e.

$$b \cdot 2Z > 4a\{(A - 2Z - 1) + 2\}.$$

These conditions may be expressed in terms of an allowable variation of the function $b \cdot 2Z/4a$ as:

for (A, Z) stable, the range is $0 \to 1 + \dfrac{N}{Z - 1}$ from the

value $(A - 2Z - 1)$,

for first neutron, the range is $1 \to 1 + \dfrac{N}{Z - 1}$ from the

value $(A - 2Z - 1)$,

for second neutron, the range is $2 \to 1 + \dfrac{N}{Z - 1}$ from the

value $(A - 2Z - 1)$.

Since N/Z varies from 1 for light nuclei to $\sim 1\frac{1}{2}$ for the heaviest nuclei, these ranges are, for the heavy nuclei $2\frac{1}{2}$ units, $1\frac{1}{2}$ units, and $\frac{1}{2}$ unit respectively. Thus the probabililty of a neutron adding is roughly $\frac{3}{5}$, and a further neutron roughly $\frac{1}{3}$. A similar simple argument with protons would lead to $\frac{2}{5}$ for the first proton and zero for the second. Using all the terms of the mass equation and the more exact functions for the two discussed does not alter the general conclusion that if (A, Z) is stable then the addition of two more nucleons is more likely to be of the form (p, n) for stability than (2p) or (2n). Thus odd–odd nuclei should be as prolific as even–even.

Obviously a term is missing from the mass equation. It must be such that if A and Z are both even then, whilst not altering the probability that the first particle be a neutron for stability ($\frac{3}{5}$ is just about correct for a heavy nucleus), it must increase the probability for a second neutron giving stability up to unity. This extra term is not needed in considering the stability of odd-A nuclei, but must be chosen such that in going from even A to $A + 2$, where both nuclei are stable, pairs of like particles will be added. Hence the form adopted.

Appendix C: Exchange forces

To introduce this concept it is interesting to consider the singly ionized H_2 molecule—a system of two protons and one electron. When the separation $|R| = |r_2 - r_1|$ is very large the electron wave function will look like the hydrogen-atom bound state situated at r_1 or r_2 and designated by $\phi(r_3 - r_1)$ or $\phi(r_3 - r_2)$. However, the correct wave function must be symmetric (or antisymmetric) with respect to the two protons. Therefore, put $\psi = \alpha_\pm^{-\frac{1}{2}}\{\phi(r_3 - r_1) \pm \phi(r_3 - r_2)\}$, where the normalization integral α takes the value 2 as $R \to \infty$, whilst $\alpha_+ \to 4$, $\alpha_- \to 0$ as $R \to 0$. The energy integral is

$$4\pi\varepsilon_0 \frac{V}{e^2} - \frac{1}{R} = -\int \psi^* \left(\frac{1}{|r_3 - r_1|} + \frac{1}{|r_3 - r_2|} \right) \psi \, d\tau_3$$

$$= -2 \int \frac{\psi^*\psi}{|r_3 - r_1|} d\tau_3,$$

by symmetry. Substituting for ψ gives four integrals, and for ϕ real (corresponding to a bound state) two are the same giving (where $r = r_3 - r_1$)

$$4\pi\varepsilon_0 \frac{V}{e^2} - \frac{1}{R} = -2\alpha_\pm^{-1} \left\{ \int \phi^2(r) \frac{1}{r} d\tau + \int \phi^2(r) \frac{1}{|r - R|} d\tau \right.$$

$$\left. \pm 2 \int \phi(r - R)\phi(r) \frac{1}{r} d\tau \right\}. \tag{C.1}$$

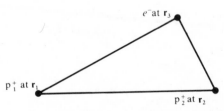

Fig. C.1.   Schematic representation of the hydrogen-molecule ion.

The first term represents the binding energy of a hydrogen atom, the second term represents the electronic contribution to the Coulomb energy of a proton distant R from a hydrogen atom; for large R this term exactly cancels the proton repulsion ($1/R$ on the left-hand-side), but for small R it becomes the finite binding energy of the hydrogen atom. The last term, unlike the other two, has no classical analogue; it effectively arises from not knowing to which proton the electron belongs. Its sign depends upon the overlap of the two atomic wave functions when separated by R and, if the wave functions have nodes, can be positive or negative. For the lowest s-state there are no nodes, so the integral is always positive, giving increased binding for the symmetric wave function and reduced for the antisymmetric. At large distances bound wave functions fall off exponentially. If the s-state looks like $\exp(-\kappa r)$ at large r, the main contribution to the integral will occur nearly halfway between the two protons and will obviously contain $\exp(-\kappa R)$ in the product of the wave functions (this qualitative argument neglects the effects of terms like $1/r$ in the integral). Thus the integral will also behave like $\exp(-\kappa R)$ at large distances.

Now assume that the existence of the electron is not known, but that experiments can be performed to determine the potential between a proton (charged) and a hydrogen atom (neutral) as a function of separation R. The classical term will be constant at large R (but unknown since the structure of hydrogen is unknown); however, at small R, the proton gets inside the electron cloud and experiences a net

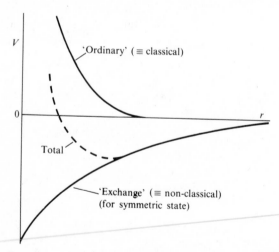

Fig. C.2. Potential of hydrogen-molecule ion. $V = 0$ corresponds to infinite separation between ionized atom (p) and neutral atom (H).

repulsion by the other proton. For the symmetric case, the non-classical term increases exponentially (approximately) to a finite binding at $R = 0$. The state of affairs is presented graphically in Fig. C.2.

Thus there exists a potential between the proton and the hydrogen atom which depends upon the spatial symmetry of the (p, H) system. During the interaction the proton and the hydrogen atoms cannot be individually identified, and this is interpreted as exchange of charge and spin (in the form of an unknown electron). From Fig. C.2 this can result in attraction at large R, whilst the repulsion at small R arises because the 'fundamental-particle' hydrogen atom in fact has a structure.

The analogue with nucleons is obvious, at least for the (p, n) system. A similar argument to the above can be applied if n is looked upon as $p + \pi^-$ and gives the correct range for the nuclear force. Since the proton and neutron have almost the same mass, the total energy (rest + binding) of the bound π^- must be nearly zero.† But $\kappa = |p|/\hbar$ in the unphysical region, where $|p|$ is given by $p^2c^2 = E^2 - m_\pi^2 c^4$ with $E = 0$. Therefore $\kappa = m_\pi c/\hbar = (1.4 \text{ fm})^{-1}$. In order to produce charge independence, a π^0 must be introduced (to give the (p, p) and (n, n) interactions) and also a π^+ (though its need is not immediately obvious; it arises because the above picture has made the proton look 'more fundamental' than the neutron—the π^+ redresses the balance since its exchange appears to make the proton look like $n + \pi^+$).

Having gone through the above reasoning, it is convenient to forget the mechanism and express the different exchanges mathematically. The operator $s_1 \cdot s_2 \equiv \frac{1}{2}\{(s_1 + s_2)^2 - s_1^2 - s_2^2\} = \frac{1}{2}\{S^2 - s_1^2 - s_2^2\}$ takes the value $\frac{1}{2}\{S(S + 1) - s_1(s_1 + 1) - s_2(s_2 + 1)\}$ for the spin state S, resulting from the combination of two spin states s_1 and s_2. For $s_1 = s_2 = \frac{1}{2}$, it takes the value $+\frac{1}{4}$ for $S = 1$ and $-\frac{3}{4}$ for $S = 0$. Thus the operator $P^\sigma \equiv \frac{1}{2} + 2s_1 \cdot s_2 = 1$ for $S = 1$, triplet state, and $= -1$ for $S = 0$, singlet state. This P^σ is the spin-exchange operator. If ↑↓ is used to denote the state in which particle 1 has spin up, particle 2 spin down, then it can be represented thus:

$$\uparrow\downarrow = \tfrac{1}{2}(\uparrow\downarrow + \downarrow\uparrow) + \tfrac{1}{2}(\uparrow\downarrow - \downarrow\uparrow) = \frac{1}{\sqrt{2}}\{(S = 1) + (S = 0)\},$$

therefore

$$P^\sigma \uparrow\downarrow = \frac{1}{\sqrt{2}}\{(S = 1) - (S = 0)\} = \tfrac{1}{2}(\uparrow\downarrow + \downarrow\uparrow) - \tfrac{1}{2}(\uparrow\downarrow - \downarrow\uparrow) = \downarrow\uparrow.$$

† The unknown force binding pion and nucleon must be much stronger than the Coulomb force. Even if it has effectively zero range, the integrals will result in a nucleon–nucleon interaction with the correct range.

TABLE C.1

ψ	Symmetry‡			Sign of operator†			
	Spin	Isospin	Space	W	H	B	M
$^3S_1$	s	a	s	+	−	+	+
$^1S_0$	a	s	s	+	+	−	+
$^3P_{012}$	s	s	a	+	+	+	−
$^1P_1$	a	a	a	+	−	−	−

† See p. 32 for definition of W,H,B,M.
‡ s denotes 'symmetric'; a denotes 'antisymmetric'.

Thus P^σ has exchanged spins. P^T, which exchanges charge, can be defined similarly. The general potential is then (see Chapter 2, p. 32, for nomenclature) $V = V_W + V_H P^T + V_B P^\sigma + V_M P^S$ (where $P^S = -P^T P^\sigma$ is a requirement of the Pauli principle). Since the exchange operator has a purely numerical part, the potential can be rewritten

$$V = V_1 + V_2 t_1 \cdot t_2 + V_3 s_1 \cdot s_2 + V_4(s_1 \cdot s_2)(t_1 \cdot t_2)$$

and the operators are (loosely) referred to as exchange operators.

The effect of the operators, P^σ, P^T, P^S, is shown in Table C.1 for S- and P- states. Notice that it is the pattern of the changes of sign which is important. The sign of a whole column can be changed merely by changing the sign of the potential function accompanying the operator.

Appendix D: Spin matrices

Angular momentum can be introduced into quantum mechanics by inserting the operator form $p = -i\hbar\nabla$ into the classical expression $J = r \wedge p$, to give $J = -i\hbar(r \wedge \nabla)$. It is left to the reader to show that $J \wedge J = i\hbar J$ (hint: use Cartesian coordinates), i.e. $M_x M_y - M_y M_x = i\hbar M_z$ and two others, cyclically.

Although these equations have been derived from considerations of orbital angular momentum which is quantized to integer values, Dirac has used them to show that odd half-integer values are also permitted indicating the existence of another type of intrinsic angular momentum. Here it is proposed to consider $J = \frac{1}{2}$, and to designate it s, the intrinsic spin angular momentum of a fundamental particle. For this particular case it can be shown that $s_x s_y + s_y s_x = 0$, etc. and therefore $s_x s_y = (i\hbar/2)s_z$, etc. Since s_z is two-valued, having values $\pm\frac{1}{2}\hbar$, the wave function representing an arbitrary state of spin can be written as a two-component vector $\begin{bmatrix} a \\ b \end{bmatrix}$ in an appropriate two-dimensional space. The components of s, namely, s_x, s_y, s_z will be represented by 2×2 matrices, of which s_z will be diagonal and of the form $\frac{1}{2}\hbar\begin{bmatrix} 1 & 0 \\ 0 & -1 \end{bmatrix}$ to give the required eigenvalues when acting on the eigenfunctions $\begin{bmatrix} 1 \\ 0 \end{bmatrix}$ and $\begin{bmatrix} 0 \\ 1 \end{bmatrix}$ representing 'spin up' and 'spin down'. Putting $s_x \equiv \begin{bmatrix} a & b \\ c & d \end{bmatrix}$, $s_y \equiv \begin{bmatrix} e & f \\ g & h \end{bmatrix}$, and using $s_y s_z = \frac{1}{2}i\hbar s_x$ and $s_z s_x = \frac{1}{2}i\hbar s_y$ results in

$$a = e = d = h = 0, \; g = ic, \; f = -ib.$$

There is some latitude in the choice of s_x, s_y; the forms commonly chosen are

$$s_x = \frac{1}{2}\hbar\begin{bmatrix} 0 & 1 \\ 1 & 0 \end{bmatrix} \quad \text{and} \quad s_y = \frac{1}{2}\hbar\begin{bmatrix} 0 & -i \\ i & 0 \end{bmatrix}.$$

Notice that $s_x^2 + s_y^2 + s_z^2 = s^2 = \frac{3}{4}\hbar^2$, which conforms to the usual expression $J^2 = \hbar^2 J(J+1)$.

The raising operator

$$\hbar^{-1}s^+ \equiv \hbar^{-1}(s_x + is_y) = \begin{bmatrix} 0 & 1 \\ 0 & 0 \end{bmatrix}$$

obviously operates on $\begin{bmatrix} 1 \\ 0 \end{bmatrix}$ to give $\begin{bmatrix} 0 \\ 0 \end{bmatrix}$ and on $\begin{bmatrix} 0 \\ 1 \end{bmatrix}$ to give $\begin{bmatrix} 1 \\ 0 \end{bmatrix}$; hence its name. Similarly the lowering operator $\hbar^{-1}s^-$ is defined as $\hbar^{-1}(s_x - is_y)$.

In a similar way the components of isospin may be defined. As for spin, the two-component isospin space has no connection with any classical space, but, and in contrast to spin, neither does the three-component isospin space in which the components τ_x, τ_y, τ_z are defined. There is a one-to-one correspondence between the z-component of isospin and the charge of the nucleon, but otherwise there is no further definition of this space. The isospin raising and lowering operators have the property of changing one nucleon type to its counterpart, without changing any other part of its wave function. It should be stated that this is a rather restricted view of isospin since only nucleons are under consideration as distinct from its greater ramifications when dealing with other hadrons.

Appendix E: Magnetic moments of nuclei

The magnetic moment vector of a single particle in the nucleus is given by $\mu_N(g_l\mathbf{l} + g_s\mathbf{s})$ where μ_N is the nuclear magneton $(e\hbar/2M_p)$ and g_l and g_s are the respective gyromagnetic ratios for orbital and spin angular momentum reflecting the fact that magnetism arises from circulating (or spinning) charges. The proton has $g_l = 1$ and $g_s = +5.58$ whilst the neutron has $g_l = 0$ and $g_s = -3.83$ ($g_l = 0$ is expected for an uncharged particle but g_s need not be zero since there could be a charge distribution within the neutron—a classical sphere with a positive charge distribution near the centre turning into a negative charge distribution near the surface could have a nett charge of zero but would generate a magnetic field when set spinning). The point has been noted elsewhere in the book that for a true 'point' charge the Dirac wave equation would give $|g_s| = 2$. That $g_s \neq 2$ for the proton and $g_s \neq 0$ for the neutron must be ascribed to structures for these particles—for example they can generate meson fields around themselves. They are also known to have considerable structure from electron scattering at extremely high energies, resulting in the Quark Theory of fundamental particles—a proton consists of three basic quarks interacting with each other and the interactions themselves can be looked upon in terms of fields of other virtual particles, possibly charged, so the combined structure can have a complicated charge distribution.

The energy of interaction of a nucleus with a magnetic field B will therefore be given by

$$\langle \psi_0^* | -\mu_N \sum_i (g_{l_i}\mathbf{l}_i + g_{s_i}\mathbf{s}_i) \cdot \mathbf{B} | \psi_0 \rangle$$

where ψ_0 is the ground-state wave function and the summation is taken over all particles within the nucleus. In general ψ_0 is not sufficiently well known to allow us to determine this expression. Recourse must be made to simplifying models for example the extreme single particle shell model (see Chapter 3) which is perhaps too much of a simplification and was rapidly superseded by the general multi-particle shell model. In the former the assumption is made that protons pair off with each other to give zero angular momentum for all Z protons if Z is even and for $(Z-1)$ protons if Z is odd, with the highest lying proton unpaired; and that neutrons behave similarly. Thus all even–even nuclei will have $J = 0$ and therefore no magnetic interaction (this is the case for all even–even ground states which gives some validity to the model) whilst odd–even nuclei have J given by j ($= l + s$) of the odd particle, which is unpaired; odd–odd nuclei must have an odd proton and an odd neutron and are more properly dealt with in the more general shell model treatment which attempts to cope with the coupling together of all the nucleons outside closed shells. Here we shall limit our discussions to the extreme single-particle treatment of odd nuclei with a brief extension to odd–odd nuclei.

Odd nuclei

For a single particle we therefore drop the summation in our expression for the interaction energy and arrive at four different formulae corresponding to whether the particle is a proton or a neutron with each of these cases further split according as $j = l + \frac{1}{2}$ or $l - \frac{1}{2}$. The basic treatment can be simply described in terms of the vector model; the vectors l and s interact to form j and the system precesses around j with a frequency equal to the interaction energy divided by h. When an external field B is applied then j will precess around B again with a frequency equal to this interaction energy divided by h. If the former frequency is very much higher than the latter the contribution of the orbital motion to the magnetic energy must be evaluated first by averaging $g_l l$ along j. The classical average will be $g_l(l \cdot j/j^2)j$ and $l \cdot j$ is obtained from $s = l - j$, therefore $s^2 = (l - j)^2 = l^2 + j^2 - 2l \cdot j$ giving $l \cdot j = \frac{1}{2}(l^2 + j^2 - s^2)$. Quantum mechanically, under the condition imposed, j^2, l^2, and s^2 are constants of the motion having the values $j(j+1)$, $l(l+1)$, and $s(s+1)$ so

$$\langle g_l l \rangle = g_l \frac{j(j+1) + l(l+1) - s(s+1)}{2j(j+1)} j$$

and a similar expression obtains for $\langle g_s s \rangle$ by interchanging l and s in the expression given. The total energy of interaction is therefore

$$-\mu_N \left[g_l \frac{j(j+1) + l(l+1) - s(s+1)}{2j(j+1)} + g_s \frac{j(j+1) + s(s+1) - l(l+1)}{2j(j+1)} \right] j \cdot B$$

which splits the degenerate ground state into $(2j+1)$ levels as $j \cdot B$ takes on the values $m_j B$ where m_j goes from $-j$ to $+j$ in unit steps. As has been stated in Chapter 1, the 'scalar' magnetic moment μ is defined in terms of its greatest alignment along B giving $\mu = \mu_N g j$ where g is the nuclear g-factor and is the bracketed term in the above expression for the magnetic energy. As defined it is not just a scalar magnitude since it also has a sign ascribed to it—the sign of g—which is positive when the magnetic moment vector is directed along j and negative when in the reverse direction.

Putting in the four different cases we have:

Odd proton

(a) $j = l + \frac{1}{2}$, $\mu = (j - \frac{1}{2} + \frac{1}{2}g_{sp})\mu_N = (j + 2.29)\mu_N$;

(b) $j = l - \frac{1}{2}$, $\mu = j(j + 1)^{-1}(j + 3/2 - \frac{1}{2}g_{sp})\mu_N = j(j + 1)^{-1}(j - 1.29)\mu_N$.

Odd neutron

(a) $j = l + \frac{1}{2}$, $\mu = \frac{1}{2}g_{sn}\mu_N = -1.91\mu_N$;

(b) $j = l - \frac{1}{2}$, $\mu = -j(j + 1)^{-1}\frac{1}{2}g_{sn}\mu_N = j(j + 1)^{-1}1.91\mu_N$.

These expressions plotted against j are known as the Schmidt lines, and are given in most textbooks. When the measured values for various odd nuclei are filled into the respective plots it is found that they do not lie very close to their appropriate Schmidt lines, which is hardly surprising since the strict single particle shell model is a gross over-simplication. The groupings of the measured values however show a strong connection with the Schmidt lines, and, as mentioned in Chapter 3, a number of nuclei of the type 'closed shell $\pm$ one nucleon' fall very close to the appropriate points on these lines indicating that their ground state wave functions are dominated by the single particle, or hole, structure. It is intriguing to note that the measured values fall between the two Schmidt lines with few exceptions. The explanation of this, which is not simple, arises from the fact that nuclei away from closed shells are deformed into ellipsoids. In such a deformed potential well, the nuclear wave function for a state thought to be of the type $j = l + \frac{1}{2}$ can have mixed into it a component having $j = l - \frac{1}{2}$ coupled with nuclear rotation. The two very simple nuclei ^{3}He and ^{3}H fall outside the Schmidt lines, but close to their appropriate positions; it is thought that the meson cloud is responsible for the small discrepancies, especially since they are roughly equal and opposite in sign—this should be the case for the meson cloud which will tend to be negative for the odd neutron and positive for the odd proton.

An odd–odd nucleus

As an example take $^6_3$Li consisting of an odd proton and an odd neutron outside the closed shell α-particle core with the ground state J^π of 1^+. How to combine their effects depends on how they are coupled together.

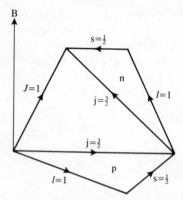

Fig. E.1. Vector model of j–j coupling of two $p_{3/2}$ particles to $J = 1$.

To begin with we will assume that they are both $p_{\frac{3}{2}}$ particles and are j–j coupled which is in keeping with the shell model. The vector model for the coupling is given in Fig. E.1. The proton $\boldsymbol{l}$ and $\boldsymbol{s}$ magnetic terms must be projected onto $\boldsymbol{j}_p$ and similarly the neutron $\boldsymbol{l}$ and $\boldsymbol{s}$ magnetic terms must be projected onto $\boldsymbol{j}_n$; then these two projections, now along $\boldsymbol{j}_p$ and $\boldsymbol{j}_n$, must be projected on to $\boldsymbol{J}$. Thus

$$\boldsymbol{\mu}_p = \mu_N \left(g_{l_p} \frac{\boldsymbol{l}_p \cdot \boldsymbol{j}_p}{j_p^2} + g_{s_p} \frac{\boldsymbol{s}_p \cdot \boldsymbol{j}_p}{j_p^2} \right) \boldsymbol{j}_p$$

and

$$\boldsymbol{\mu}_n = \mu_N \left(g_{l_n} \frac{\boldsymbol{l}_n \cdot \boldsymbol{j}_n}{j_n^2} + g_{s_n} \frac{\boldsymbol{s}_n \cdot \boldsymbol{j}_n}{j_n^2} \right) \boldsymbol{j}_n$$

and further projecting onto $\boldsymbol{J}$:

$$\boldsymbol{\mu} = \mu_N \left[\left(g_{l_p} \frac{\boldsymbol{l}_p \cdot \boldsymbol{j}_p}{j_p^2} + g_{s_p} \frac{\boldsymbol{s}_p \cdot \boldsymbol{j}_p}{j_p^2} \right) \frac{\boldsymbol{j}_p \cdot \boldsymbol{J}}{J^2} + \left(g_{l_n} \frac{\boldsymbol{l}_n \cdot \boldsymbol{j}_n}{j_n^2} + g_{s_n} \frac{\boldsymbol{s}_n \cdot \boldsymbol{j}_n}{j_n^2} \right) \frac{\boldsymbol{j}_n \cdot \boldsymbol{J}}{J^2} \right] \boldsymbol{J}.$$

But

$$\frac{\boldsymbol{l}_p \cdot \boldsymbol{j}_p}{j_p^2} = \frac{j_p(j_p + 1) + l_p(l_p + 1) - s_p(s_p + 1)}{2 j_p (j_p + 1)}$$

$$= \frac{\frac{3}{2} \cdot \frac{5}{2} + 1 \cdot 2 - \frac{1}{2} \cdot \frac{3}{2}}{2 \cdot \frac{3}{2} \cdot \frac{5}{2}} = \frac{2}{3}$$

$$\frac{\boldsymbol{s}_p \cdot \boldsymbol{j}_p}{j_p^2} = \frac{j_p(j_p + 1) - l_p(l_p + 1) + s_p(s_p + 1)}{2 j_p (j_p + 1)}$$

$$= \frac{\frac{3}{2} \cdot \frac{5}{2} - 1 \cdot 2 + \frac{1}{2} \cdot \frac{3}{2}}{2 \cdot \frac{3}{2} \cdot \frac{5}{2}} = \frac{1}{3}$$

and the two neutron projections are exactly the same. Also,

$$\frac{j_p \cdot \mathbf{J}}{J^2} = \frac{J(J+1) + j_p(j_p+1) - j_n(j_n+1)}{2J(J+1)}$$

$$= \frac{1 \cdot 2 + \frac{3}{2} \cdot \frac{5}{2} - \frac{3}{2} \cdot \frac{5}{2}}{2 \cdot 1 \cdot 2} = \frac{1}{2}$$

and the same for the neutron projection. Putting in the four g-values gives

$$\boldsymbol{\mu} = \mu_N[(1 \times \tfrac{2}{3} + 5.58 \times \tfrac{1}{3})\tfrac{1}{2} + (0 - 3.83 \times \tfrac{1}{3})\tfrac{1}{2}]\mathbf{J} = +0.625\mu_N\mathbf{J}$$

and since M_J maximum is 1, $\mu = +0.625\mu_N$. The measured value for ^{6}Li is $+0.822\mu_N$, which, although fairly close, may be an indication that the j–j coupling scheme is not correct. If we assume L–S coupling and add the two orbital momenta to $L = 0$ and the two spins to $S = 1$ then we need to project only the spin magnetic moments to get a result. This is obviously equivalent to the calculation for the magnetic moment of the deuteron in the pure $^3S_1$ ground state which gives the result $\mu = +0.88\mu_N$—a much closer result and in keeping with the belief that, for light nuclei up to $A \sim 10$, L–S coupling occurs, with j–j coupling for heavy nuclei and in the region A just greater than 10 the coupling is neither the one nor the other but 'intermediate'.

Appendix F: Isospin in electromagnetic transitions

Electric dipole transitions

The E1 matrix element is of the form $\langle \psi_f^* | e \sum_i r_i | \psi_i \rangle$ where the summation is taken over all the protons in the nucleus. This sum can be extended to all nucleons by inserting the factor $(\frac{1}{2} - \tau_{zi})$ into the summation since this takes the value 1 for a proton and 0 for a neutron. This factor operates on the isospin component of the wave function; since the two terms in the bracket are respectively (i) a scalar and (ii) the third component of a vector in isospin space, these two parts of the matrix element will be non-zero if (i) $\Delta T = 0$ and (ii) $\Delta T = 0, \pm 1$ with $0 \to 0$ excluded. We have therefore the selection rule $\Delta T = 0, \pm 1$ for E1 radiation—in fact the argument so far holds for all electric multipoles since in all of them summation is taken over protons only.

However, the form of the matrix element given above is not strictly accurate since $e \sum_i r_i(t)$ would not be zero if the nucleus as a whole were to oscillate in space. This motion has physical reality and represents the process of Thompson scattering of radiation (the nuclear equivalent of Rayleigh scattering by electrons), but it does not produce any change in the internal structure of the nucleus and cannot therefore produce a nuclear excitation or de-excitation. To eliminate this process from consideration, the operator must be modified such that the centre of mass of the nucleus as a whole does not change.

Consider the case of a proton in the nucleus (A, Z) moving relative to the rest of the nucleus $(A - 1, Z - 1)$ by a displacement r, then in the centre of mass system, the proton moves r_p and the rest of the nucleus recoils $(r_p - r)$ such that $(A - 1)(r - r_p) = 1 \cdot r_p$ giving $r_p = r(A - 1)/A$. But in moving through r_p, the proton changes the dipole moment by $+er_p$, while the recoil core changes it by $-(Z - 1)e(r - r_p)$. The total change of dipole moment is therefore

$$er_p - (Z - 1)e(r - r_p) = e\left(\frac{A - 1}{A} - \frac{Z - 1}{A}\right)r = \frac{A - Z}{A}er.$$

For a neutron displaced through r relative to the rest of the nucleus, r_n will be $r(A-1)/A$ as for the proton since neutron and proton have the same mass. But the rest of the nucleus is now $(A-1, Z)$ and the dipole created is $0 - Ze(r - r_n) = (Z/A)er$. Thus to correct our initial matrix element we must assign effective charges of $+e(A-Z)/A$ and $-eZ/A$ to protons and neutrons respectively and this is achieved by amending the factor $(\frac{1}{2} - \tau_{zi})$ to $((A - 2Z)/2A - \tau_{zi})$. If we now look at the particular case of self-conjugate nuclei, having $N = Z = \frac{1}{2}A$, then we see that the above factor has become $-\tau_{zi}$ and the isoscalar part of the transition has vanished. To consider how the isovector term behaves let us assume that we have a nucleus consisting of a neutron and proton outside a core consisting of equal numbers of neutrons and protons forming closed shells, then the wave functions of the two 'valence' particles can be represented in isospin space as $2^{-\frac{1}{2}}(\Uparrow\Downarrow + \Downarrow\Uparrow)$ for $(T, T_z) = (1, 0)$ and $2^{-\frac{1}{2}}(\Uparrow\Downarrow - \Downarrow\Uparrow)$ for $(T, T_z) = (0, 0)$—see Appendix C. Since $\tau_z \uparrow = +\frac{1}{2}\uparrow$ and $\tau_z \downarrow = -\frac{1}{2}\downarrow$ then $\tau_{z1}(\Uparrow\Downarrow + \Downarrow\Uparrow) = \frac{1}{2}(\Uparrow\Downarrow - \Downarrow\Uparrow)$ where the second subscript is the number of the particle and when it is 1 indicates that it operates only on the left hand member of a pair of arrows. Similarly $\tau_{z1}(\Uparrow\Downarrow - \Downarrow\Uparrow) = \frac{1}{2}(\Uparrow\Downarrow + \Downarrow\Uparrow)$ whilst $\tau_{z2}(\Uparrow\Downarrow + \Downarrow\Uparrow) = \frac{1}{2}(-\Uparrow\Downarrow + \Downarrow\Uparrow)$ and $\tau_{z2}(\Uparrow\Downarrow - \Downarrow\Uparrow) = \frac{1}{2}(-\Uparrow\Downarrow - \Downarrow\Uparrow)$. Thus both terms in the sum change isospin from 1 to 0, or vice versa, so the matrix element can be non-zero only for $\Delta T = \pm 1$ transitions (but note that, although $\tau_{z1}(1, 0) = \frac{1}{2}(0, 0)$ and $\tau_{z2}(1, 0) = -\frac{1}{2}(0, 0)$ cancellation does not necessarily occur on summing over the two particles since in the matrix element τ_{z1} is associated with r_1 and τ_{z2} with r_2 giving a spatial contribution proportional to $(r_1 - r_2)$). Hence the isospin selection rule in self-conjugate nuclei: electric dipole transitions are forbidden unless the isospin changes by one unit. Note that this result is specific to E1, since the centre of mass corrections to higher multipoles are quite small, as the student can verify for himself by an extension of the argument given. Note also that for non-self-conjugate nuclei, i.e. $T_z \neq 0$, the isoscalar term is not zero although it may be small, but in any case τ_z operating on $(T, T_z \neq 0)$ will in general produce all three states T, $T \pm 1$ assuming that $T - 1 \geq T_3$, otherwise just T and $T + 1$; there is therefore no further restriction beyond that given in the opening paragraph.

Magnetic dipole transitions

A similar suppression occurs for magnetic dipole transitions, but the argument is messy and 'hand waving'. In this case the operator for the transition can be written as

$$\sum_i [(\tfrac{1}{2} - \tau_{zi})l_i + (\tfrac{1}{2} - \tau_{zi})g_p s_i + (\tfrac{1}{2} + \tau_{zi})g_n s_i]$$

in units of the nuclear magneton, with $g_p = +5.58$, $g_n = -3.83$. Again we have an isoscalar term $\sum_i [\frac{1}{2} l_i + \frac{1}{2}(g_p + g_n)s_i]$ and an isovector term $-\sum_i \tau_{zi}[l_i + (g_p - g_n)s_i]$. The isoscalar part can be written as $\frac{1}{2}J + \sum_i \frac{1}{2}(g_p + g_n - 1)s_i$ using $\sum_i l_i = J - \sum_i s_i$. Since J is a constant of the motion it has no off-diagonal matrix elements and so drops out. The transition rate is therefore proportional to $\frac{1}{4}(g_p + g_n - 1)^2$ times the square of the matrix element of $\sum_i s_i$. The isovector term cannot be so manipulated since, as seen previously for self-conjugate nuclei, τ_{z1} acting on $(1, 0)$ gives $+\frac{1}{2}(0, 0)$ but τ_{z2} acting on $(1, 0)$ gives $-\frac{1}{2}(0, 0)$, so the summation is effectively differencing the orbital angular momenta of the two particles. For a rough evaluation let us assume that this differencing process makes this orbital contribution small then the transition rate is proportional to $(g_p - g_n)^2$ times the square of the matrix element of $\sum_i \tau_{zi} s_i$. To continue this 'hand waving' argument we assert that, on average, the squares of the matrix elements are the same except for a factor $\frac{1}{4}$ in the second case arising from $\tau_z = \pm\frac{1}{2}$. The transition rates are therefore expected to be roughly in the ratio $(g_p + g_n - 1)^2 : (g_p - g_n)^2 = 0.75^2 : 9.41^2$ or about $1 : 150$.

To revert to self-conjugate nuclei, the isoscalar term is connected only with $\Delta T = 0$ transitions and the isovector term only with $|\Delta T| = 1$, leading to the selection rule for magnetic dipole transitions: in self-conjugate nuclei $\Delta T = 0$ transitions are suppressed by a factor $\sim 1/150$ relative to $|\Delta T| = 1$ transitions. Note that for non-self-conjugate nuclei the isovector term can contribute to $\Delta T = 0$ as well as to $|\Delta T| = 1$, so there is no significant difference between the two types of transition.

Validity of the selection rules

Thus we have arrived at selection rules for E1 and M1 transitions in self-conjugate nuclei. For E1 the calculation is 'clean' and would be exact if isospin were an exact quantum number—at least in the simplifying limit that the wavelength of the transition involved is large compared with the nuclear size. The M1 rule is however based on a messy calculation carried out extremely approximately and leads to a suppression factor rather than a clean selection rule. However the Coulomb force, between protons only, mixes states of different isospin to a small extent—the order of a few per cent in intensity or about 10–20 per cent in amplitude—so the clean selection rule for E1 ends up as a suppression factor of about $1/100$. The nett result is that, in self-conjugate nuclei both E1 and M1 transitions between states of the same isospin ($0 \rightarrow 0$ or $1 \rightarrow 1$) are suppressed by a factor of about 100 relative to isospin changing transitions ($1 \to 0$ or $0 \to 1$). Note also that in the more general case, the isovector term cannot change isospin by two units or more, so all electromagnetic radiations are forbidden for $|\Delta T| \geqslant 2$. Perhaps it would be more accurate to say that for $|\Delta T| = 2$ a suppression factor of order 100 arises because of isospin mixing.

Appendix G: Averaged cross-sections and isobaric analogue states

The significance of Γ/D

It is interesting to look at the behaviour of averaged total neutron cross-sections, which smooth out the resonant fluctuations and so should go over smoothly to the measured cross-sections at higher energies when the neutron energy spread achieves the averaging. Now, by definition,

$$\sigma_{av}(E) = \frac{1}{\Delta E} \int_{E - \frac{1}{2}\Delta E}^{E + \frac{1}{2}\Delta E} \sigma \, dE.$$

If ΔE is large compared with the average spacing, D, between the resonances which, for slow neutrons, is itself greater than the total width of a resonance, then there are $\Delta E/D$ resonances in the range of interest and each gives

$$\int_{-\infty}^{+\infty} \sigma(BW) \, dE = \int_{-\infty}^{\infty} \pi \lambdabar^2 S \frac{\Gamma_n \Gamma_{tot} \, dE}{(E - E_r)^2 + \Gamma_{tot}^2/4} = 2\pi^2 \lambdabar^2 S \Gamma_n,$$

since for $\Delta E \gg \Gamma_{tot}$ no great error is incurred by extending the resonance integral limits indefinitely. So, for the total neutron cross-section, $\sigma_{av} = \frac{1}{\Delta E} \cdot \frac{\Delta E}{D} \cdot 2\pi^2 \lambdabar^2 S \Gamma_n = 2\pi^2 \lambdabar^2 S \Gamma_n/D$. Obviously the factor Γ_n/D should be looked upon as an average value since Γ_n will fluctuate from resonance to resonance, and resonances are not evenly spaced. We are chiefly interested in neutrons of low energy, so resonances will make a substantial contribution to σ only if $l = 0$, see p. 42. If also we restrict ourselves to even–even target nuclei then the statistical factor S is unity since only states with $J^\pi = \frac{1}{2}^+$ produce observable resonances. The spacing D therefore refers also to states with $J^\pi = \frac{1}{2}^+$—D is always defined for specific J^π at higher energies when other J^π contribute to the yield since little progress can be made theoretically using a D defined for all resonances lumped together. With the above restrictions therefore $\sigma_{av} = 2\pi^2 \lambdabar^2 \Gamma_n/D$.

From p. 76 (adapting from α-particles to neutrons) we find that, for a single particle neutron resonance $\Gamma_n = (\hbar^2/2mR^2)(KRT)$. Moreover single particle levels occur when $KR = n\pi$ i.e. when $2mER^2/\hbar^2 = n^2\pi^2$ giving $\Delta E/\Delta n = 2n\pi^2\hbar^2/2mR^2$, so putting $\Delta n = 1$ gives $\Delta E = D$ and $D = (\hbar^2/2mR^2) \cdot 2n\pi^2 = (\hbar^2/2mR^2) \cdot 2\pi KR$ therefore $(\Gamma_n/D)_{sp} = T/2\pi$. If now we put Γ_n/D for the resonances equal to $(\Gamma_n/D)_{sp}$ we get $\sigma_{av} = \pi\lambdabar^2 T_0$ where the transmission coefficient has been given the subscript $_0$ to emphasize that this result represents the contribution which $l = 0$ makes to the cross-section. But as one increases the neutron energy other l-values make contributions to σ_{av}; one might guess that the result merely becomes $\sigma_{av} = \sum_l \pi\lambdabar^2 T_l$, but this ignores the statistical factor, which we now look at more closely. Again using an even–even target, a neutron of orbital angular momentum l makes states having $J_c = l + \frac{1}{2}$ and $l - \frac{1}{2}$ which have statistical factors $(2l + 2)/2$ and $2l/2$ respectively in their resonant cross-sections. In our simple model, we assume that $(\Gamma_n/D)_{sp} = T_l/2\pi$ for both types of states and the contributions to the average cross-section are therefore $(l + 1)\pi\lambdabar^2 T_l$ and $l\pi\lambdabar^2 T_l$ which add to $(2l + 1)\pi\lambdabar^2 T_l$. At higher energies therefore we must use $\sigma_{av} = \sum_l (2l + 1)\pi\lambdabar^2 T_l$. A simple assumption enables us to sum this expression, namely let $T_l = 1$ for $l \leqslant kR$ and $T_l = 0$ for $l > kR$—this is known as the Black Nucleus approximation since we are postulating that the nucleus absorbs all neutrons whose impact parameter, classically, is less than the nuclear radius; the critical l defines classically the maximum angular momentum which a neutron can bring in. Quantum mechanics will give a similar result, but the sharpness of the cut-off becomes blurred by the wave nature of the process. For this approximation $\sigma_{av} = (kR + 1)^2 \pi\lambdabar^2 = \pi(R + \lambdabar)^2$. This result obviously gives the classical value πR^2 at high neutron energies when $\lambdabar \ll R$ and gives $\pi\lambdabar^2$ at the low energy extreme—a result which emphasizes the wave nature of the process when $\lambdabar \gg R$, but it has retained $T_0 = 1$ from the Black Nucleus approximation rather than the more correct expression $4Kk/(K + k)^2 \sim 4k/K$ when the quantum mechanical reflection at the nuclear surface is taken into account. At high energies this factor goes to unity, so conforming with the Black Nucleus picture.

The connection between Γ_n/D (actual) and Γ_n/D (single-particle)

Thus putting $\Gamma_n/D = (\Gamma_n/D)_{sp}$ gives roughly correct averaged cross-sections, but is there a physical reason for this choice? A nuclear system should contain states which look like single-particle excitations, but, and especially as the energy of excitation increases, there are vastly more states corresponding to the same total excitation energy being shared by quite a large number of nucleons. If states were pure then these multiple excitations could not emit neutrons at all since a great fraction of the available energy would need to be concentrated on a single neutron to allow this to happen. On the other hand the pure single-particle levels could only re-emit the neutron back into the entrance channel and so could add nothing to the reaction cross-section, which would be zero. But states of the same J^π become strongly mixed with each other by the nuclear forces, so the single-particle states are mixed in with all the other states of the J^π, thereby conferring on these other states the possibility of interacting with the entrance channel—they also acquire widths in all other open channels in the same way and therefore contribute to the reaction cross-section. The above is a re-statement of Bohr's Compound Nucleus hypothesis. If the mixing is strong enough the single-particle states effectively disappear and their original neutron widths are shared over all the other states; the equation $(\Gamma_n/D)_{\text{all}} = (\Gamma_n/D)_{\text{sp}}$ then expresses the fact that there can be no creation of extra neutron width by this mixing process.

An improved connection

The optical model has a similar basis, but is an improvement on this sharing of width roughly equally between all states of the same J^π. By making the absorption inside the nucleus less strong than is implied by a black nucleus, the mixing in of the single-particle properties is restricted to a limited range around the locations of the single-particle states (the connection between mixing and the imaginary term in the potential has been indicated in Chapter 3). (Γ_n/D) as a function of energy will therefore show broad peaks at the locations of the single-particle states. As has been mentioned in Chapter 3, this has been illustrated by the analysis of slow neutron resonances—not by varying the energy, since a range of many MeV is required and the resonances (to say nothing of the experimental techniques) change radically in character beyond the first few tens of keV, but by changing A of the target nuclei! The single particle peaks are determined by $KR = n\pi$ and $R \propto A^{\frac{1}{3}}$, so the variation can be achieved in R rather than K (which, because of the large well depth, changes rather slowly with neutron energy). All the measurements can be made using slow neutron techniques to locate and analyse the resonances; this has the added advantage that all the resonances have the same J^π ($\frac{1}{2}^+$ for even–even target nuclei) rendering the determination of D simple and convincing. Finally, it should be added that the above treatment of the merging of resonance cross-sections into optical model cross-sections is not just limited to neutron induced reactions, but is also applicable to reactions induced by charged particles; T_l will then automatically include the Gamow term. But it would be more difficult to test the model in this regime because of the Gamow effect; even protons need to be well above threshold to produce measurable effects and for most nuclei so well above their threshold that they are also well above the threshold for neutron emission. The resonances are therefore very broad and partly overlapping leading to difficulties both in measurement and in analysis—especially in view of the fact that protons of a few MeV can easily take in up to two units of angular momentum and the different J values produced would lead to single particle peaks in different locations.

Isobaric analogue resonances

A dramatic example of nuclear reactions explained by the mechanisms outlined in this appendix comes under the general title of Isobaric Analogue Resonances. As has been explained in Chapter 2, a nucleus (A, Z) will have, superimposed on the levels, including the ground state, having the lowest isospin namely $T = (A/2 - Z)$, a sequence of levels corresponding to $T = (A/2 - Z + 1)$, which, after adjusting for the different Coulomb energy and the proton–neutron mass difference, find complements in the levels (from the ground state) of the nucleus $(A, Z - 1)$. In particular the lowest level of this sequence should have the same structure as the ground state of $(A, Z - 1)$. Although one generally has in mind the light nuclei when making this comparison, if isospin is still a good quantum number, it should also be applicable to heavier nuclei—for example the nucleus $^{90}\text{Zr} \equiv (90, 40)$, having $T = 5$ in the levels near ground, should have as its lowest $T = 6$ state a configuration similar to that of the ground state of $^{90}\text{Y} \equiv (90, 39)$. At what energy would it be expected? The mass equation (p. 16) can be used to obtain a rough approximation by applying it to the $T = 5$ ground state and the lowest $T = 6$ state in ^{90}Zr—in principle the equation should be applied only to ground states, but this particular excited state does in fact look like a ground state. The only terms that are different for the two states are the terms in γ and δ; although the excited state is in an even–even nucleus, it has the structure of the neighbouring odd–odd nucleus, on the other hand the term in ε, the Coulomb term, is not based in detail on the structure but only on the charge and size. The symmetry term gives a difference $\gamma[A - 2(Z - 1)]^2/A - \gamma[A - 2Z]^2/A$ which can be re-expressed as $4\gamma(2T + 1)/A$ where T refers to the ground state, giving the value ~ 11.4 MeV for ^{90}Zr (for a heavy nucleus ^{208}Pb, $T = 22$ and this term has the value ~ 20 MeV); the pairing energy difference for $A = 90$ is ~ 2.3 MeV (~ 1.4 MeV for $A = 208$). The total energy difference for ^{90}Zr is therefore ~ 13.7 MeV, and for a comparison in a heavy nucleus, ~ 21 MeV in ^{208}Pb. From mass tables $^{89}\text{Y} + \text{p}$ exceeds the mass of ^{90}Zr by 8.38 MeV, so the lowest $T = 6$ state might be expected to show up as a resonance at a proton energy of $\sim 13.7 - 8.4$ or ~ 5.3 MeV; a more accurate assessment than has been given here gives

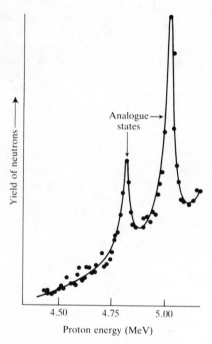

Fig. G.1.   Excitation function for the reaction ^{89}Y(p, n)^{89}Zr.

4.75 MeV (CM) or 4.80 MeV (lab). Since ^{89}Y has $T = 11/2$ in its ground state, this proton channel is able to form $T = 5$ and $T = 6$ states in the compound nucleus ^{90}Zr.

The excitation function of the reaction ^{89}Y(p, n)^{89}Zr, for the detection of all neutrons is given in Fig. G.1. It shows two resonances superimposed on a rising background; the resonances are separated by ~200 keV which compares well with the separation of 202.4 keV between the ground state and first excited state of ^{90}Y. The resonances are of the order of 30 keV wide. The rising background is what was expected since, at an excitation of order 13 MeV in the compound nucleus, the level spacing of $T = 5$ levels is probably a few eV, so the measurements average over many resonances and a large range of neutron energies which leads to smooth cross-sections with the rise reflecting the influence of the Coulomb barrier in the entrance channel. If one interprets the two resonances as the lowest $T = 6$ states, for pure T, they should have no neutron width since neutron decay should be to $T = 11/2$ in the final nucleus ^{89}Zr, which has $T = 9/2$ near ground. The lowest $T = 11/2$ states will lie some 10 or 11 MeV higher and are not accessible in the (p, n) reaction at a proton energy of 5 MeV.

The explanation of the reaction is along the lines of this appendix. The $T = 6$ states are mixed into the background $T = 5$ states over an energy range which depends upon the strength of the mixing interaction. In the previous treatment, the mixing is produced by (part of) the strong interaction and is of order of 1 MeV; here the weaker Coulomb interaction is responsible and a few tens of keV is reasonable. The sharpness of the resonances makes them easy to detect. Although the yield looks like that of a simple single-state resonance, each measurement is really an integral over hundreds or thousands of states; it so happens that these states acquire extra width in the entrance channel because of the underlying $T = 6$ states.

The study of isobaric analogue resonances has been of interest, not only because it emphasizes that isospin is a good quantum number up to the heaviest nuclei, but also because it has given information on nuclear structure not previously available. An example of this is the Coulomb energy term in the location of the resonances; one can see how the Coulomb energy changes with the structure of the state in question since one is concerned with a specific particle, not the nucleus as a whole. Concerning the 'goodness' of the isospin quantum number, before the discovery of these resonances, it was thought that the mixing of isospin became progressively stronger with increasing Z, but it was realized that this was so only as far as $A \sim 40$–50, when T is very small, but for $A > 50$ T itself progressively increases for stable nuclei and the mixing, found to be proportional to $Z^2/(2T + 1)$, stays approximately constant.

Isobaric analogue states can be excited by direct processes e.g. the (p, n) reaction in which the charge exchange term in the nuclear interaction results in the proton projectile exchanging its charge with a neutron in the target nucleus without changing the state of this bound particle. The state formed in the final nucleus is a member of the same isospin multiplet as the ground state of the initial nucleus—but again it is mixed into the background states having T one unit less and these can be excited by the compound nucleus process. The process is then observed by seeing a sharp peak on a broader background in the neutron energy spectrum.

Appendix H: Cross-sections by method of partial waves

Expansion of the plane wave

In the previous appendix continuous cross-sections were obtained from the overlap of resonances, the cross-section for a resonance having been derived from Fermi's Golden Rule. In that approach the 'scale factor' of a cross-section, $\pi \lambda^2$, arose, almost incidentally, in the process of eliminating the superfluous density of states factor in the cross-section when trying to express it symmetrically with respect to incoming and outgoing channels (more specifically, making σ a function of the product $\Gamma_{in}\Gamma_{out}$). In the method now to be discussed this factor arises as soon as we recognize that the beam must be represented by a wave rather than a stream of classical particles. Despite the fact that it would therefore seem to be the more natural approach to the topic of nuclear reactions it has not been adopted in the main chapters of this book because the mathematics underlying the equations presented in this appendix lies beyond that expected of an undergraduate in physics. A student interested in a deeper study of this topic should consult books like Schiff or Blatt and Weisskopf which often quote Whittaker and Watson's *Modern analysis* for some of the underlying mathematics.

The spherical harmonics $Y_{lm}(\theta, \varphi)$ have already been introduced to the reader as the angular dependence of the wave function arising from a spherically symmetric potential corresponding to a total orbital angular momentum $l\hbar$ and component along the z-axis of the coordinate system $m\hbar$. Mathematically they form a closed set of orthogonal functions so any spatial function can be expanded thus:

$$f(r, \theta, \varphi) = \sum_{l=0}^{\infty} \sum_{m=-l}^{+l} A_{lm}(r) Y_{lm}(\theta, \varphi).$$

Applying this to the plane wave in the entrance channel, $e^{i(kz-\omega t)}$, gives, dropping the factor $e^{-i\omega t}$

$$\psi = e^{ikz} = e^{ikr \cos \theta} = \sum_{l=0}^{\infty} (4\pi)^{\frac{1}{2}}(2l+1)^{\frac{1}{2}}i^{l}j_{l}(kr)Y_{l0}(\theta)$$

where the spherical Bessel functions $j_l(kr)$ have been introduced previously (p. 50). It should be noted that this is already a simplification of the more general expansion

$$\exp(i\boldsymbol{k} \cdot \boldsymbol{r}) = 4\pi \sum_{l=0}^{\infty} \sum_{m=-1}^{+1} i^l j_l(kr) Y_{lm}^*(\Theta, \Phi) Y_{lm}(\theta, \varphi)$$

where (Θ, Φ) represents the angular direction of the vector $\boldsymbol{k}$; put this direction $\equiv (0, 0)$ to get the previous equation.†

The spherical Bessel function $j_l(kr)$ behaves near the origin like $(kr)^l/(1 \cdot 3 \cdot 5 \cdot \cdots \cdot (2l+1))$ and as $r \to \infty$ like $\sin(kr - \frac{1}{2}l\pi)/kr$. Some examples have been sketched for a limited region including the origin in Fig. 3.2. Thus at a large distance away from the origin a plane wave moving along the direction of increasing z looks like

$$e^{ikz} = \sum_{l=0}^{\infty} (4\pi)^{\frac{1}{2}} (2l+1)^{\frac{1}{2}} i^l \frac{\sin(kr - \frac{1}{2}l\pi)}{kr} Y_{l0}(\theta)$$

$$= \sum_{l=0}^{\infty} (4\pi)^{\frac{1}{2}} (2l+1)^{\frac{1}{2}} \frac{i^l}{kr} Y_{l0}(\theta) \frac{1}{2i} [e^{i(kr - \frac{1}{2}l\pi)} - e^{-i(kr - \frac{1}{2}l\pi)}]$$

Now e^{ikr}/kr and e^{-ikr}/kr represent spherically symmetric waves diverging from and converging on the origin and correspond to the two terms in the above sum for $l = 0$. This makes it apparent that the sum itself represents waves of different l converging on and diverging from the origin and taking in or out no z-component of orbital angular momentum. Conservation of particles requires that for each l the two types of wave have the same intensity and the above expansion shows this to be so.

The scattering parameters 'at infinity'

At this stage there is nothing to distinguish the origin—it is merely a characteristic point of our coordinate system—but we shall now place a nucleus with its centre at the origin and at the same time restrict our beam to uncharged particles of spin zero. This restriction is necessary in

† For completeness, the normalized spherical harmonics are:

$$Y_{lm}(\theta, \varphi) = \frac{(-)^{l+m}}{2^l \cdot l!} \left[\frac{2l+1}{4\pi} \frac{(l-m)!}{(l+m)!} \right]^{\frac{1}{2}} (\sin\theta)^m \left(\frac{\mathrm{d}}{\mathrm{d}\cos\theta} \right)^{l+m} [(\sin\theta)^{2l}] e^{im\varphi}.$$

The first four are:

$$Y_{00} = \left(\frac{1}{4\pi} \right)^{\frac{1}{2}}, \quad Y_{11} = \left(\frac{3}{8\pi} \right)^{\frac{1}{2}} \sin\theta e^{i\varphi},$$

$$Y_{10} = \left(\frac{3}{4\pi} \right)^{\frac{1}{2}} \cos\theta, \quad Y_{1-1} = \left(\frac{3}{8\pi} \right)^{\frac{1}{2}} \sin\theta e^{-i\varphi}$$

order to retain the same expressions for the plane wave in the asymptotic limit of large r. With the nucleus at the origin interacting with the wave, causality requires that the ingoing wave does not change, but the outgoing wave can do so. The effect of the nucleus can therefore be represented by changing the bracket to $[\eta_l e^{i(kr-\frac{1}{2}l\pi)} - e^{-i(kr-\frac{1}{2}l\pi)}]$ where the modifying factor η_l is in general complex with $|\eta_l| \leqslant 1$ since particles can be absorbed but not created. It is also convenient to rewrite this as $[e^{i(kr-\frac{1}{2}l\pi)} - e^{-i(kr-\frac{1}{2}l\pi)} + (\eta_l - 1)e^{i(kr-\frac{1}{2}l\pi)}]$ and to use the first two terms to reconstitute the plane wave to give

$$\psi = e^{ikz} + \sum_{l=0}^{\infty} (4\pi)^{\frac{1}{2}}(2l+1)^{\frac{1}{2}}i^l \frac{\eta_l - 1}{2i} \frac{e^{i(kr-\frac{1}{2}l\pi)}}{kr} Y_{l0}(\theta).$$

Thus in the presence of the nucleus the plane wave is modified by the addition of diverging waves of different l. To go from here to an elastic scattering cross-section we need to evaluate the flux which is given by the expression $(-i\hbar/2m)(\psi^* \text{ grad } \psi - \psi \text{ grad } \psi^*)$; at first sight it appears that the full wave function must be used but if we look further into the physical conditions, the term e^{ikz} can be dropped when evaluating the partial cross-section at all angles except $\theta = 0$. The explanation of this is that our plane wave in practice is limited in lateral extension by an aperture of, for example, 1 mm in diameter so observations made outside the cylinder defined by the beam are determined by the rest of the wave function—to define the lateral extent of the 'plane wave' instantly distorts it by adding to it a diffraction pattern of the defining aperture, but the diffracting angle ($\sim\lambda/\text{diam}$) is so utterly small that diffraction can be neglected for any practical angle which takes the detector outside the beam itself.

The outgoing flux is therefore given by

$$\mathcal{Y}(\theta) \, d\Omega = \frac{4\pi v}{k^2} \left| \sum_{l=0}^{\infty} (2l+1)^{\frac{1}{2}} \left(\frac{\eta_l - 1}{2i} \right) Y_{l0}(\theta) \right|^2 d\Omega$$

where $d\Omega = 2\pi \sin\theta \, d\theta$ for axial symmetry. This expression follows from the fact that $(-i\hbar/2m)(\psi^* \text{ grad } \psi - \psi \text{ grad } \psi^*)$ where $\psi = (A/r)e^{ikr}$ gives $v|A|^2$ when r is so large that only the leading term in $\text{grad}(e^{ikr}/r)$ is significant. Since the plane wave has been normalized to one particle per unit volume and therefore to a flux of v particles per second per unit area, the differential cross-section for elastic scattering is given by

$$v \frac{d\sigma_{el}(\theta)}{d\Omega} d\Omega = \mathcal{Y}(\theta) \, d\Omega$$

therefore

$$\frac{d\sigma_{el}(\theta)}{d\Omega} = \frac{4\pi}{k^2} \left| \sum_{l=0}^{\infty} (2l+1)^{\frac{1}{2}} \left(\frac{\eta_l - 1}{2i} \right) Y_{l0}(\theta) \right|^2$$

and the total elastic scattering cross-section is given by

$$\sigma_{el} = \sum_{l=0}^{\infty} \frac{4\pi}{k^2}(2l+1)\left|\frac{\eta_l - 1}{2i}\right|^2 = \pi\lambda^2 \sum_{l=0}^{\infty}(2l+1)|\eta_l - 1|^2$$

since the spherical harmonics are orthogonal and normalized. A simple case of practical importance is that of no absorption i.e. the entrance channel is the only channel available. Here $|\eta_l| = 1$, in order to balance the ingoing and outgoing fluxes, and we put $\eta_l = e^{i2\delta_l}$ where δ_l is real (the factor 2 here avoids a factor $\frac{1}{2}$ later on) giving

$$\sigma_{el} = \frac{4\pi}{k^2} \sum_{l=0}^{\infty}(2l+1)\sin^2\delta_l = 4\pi\lambda^2 \sum_{l=0}^{\infty}(2l+1)\sin^2\delta_l.$$

In the case of $|\eta_l| < 1$, the rate of absorption of particles gives a measure of the total non-elastic (or absorption) cross-section. Obviously a treatment limited to the entrance channel can merely take account of the loss of particles; it cannot say anything further about the possible reactions occurring. The whole wave function must now be included in deriving the net flux and the expression to use is the original one:

$$\psi = \sum_{l=0}^{\infty}(4\pi)^{\frac{1}{2}}(2l+1)^{\frac{1}{2}}\frac{i^{(l-1)}}{2kr}Y_{l0}(\theta)[\eta_l e^{i(kr-\frac{1}{2}l\pi)} - e^{-i(kr-\frac{1}{2}l\pi)}].$$

The outgoing flux is therefore given by

$$v \sum_{l=0}^{\infty} 4\pi(2l+1)\frac{1}{4k^2}|\eta_l|^2;$$

since the outgoing flux when there is no absorption is obtained by putting $|\eta_l|^2 = 1$, then the absorbed flux is given by

$$v \sum_{l=0}^{\infty} 4\pi(2l+1)\frac{1}{4k^2}(1-|\eta_l|^2) = v \sum_{l=0}^{\infty} \pi\lambda^2(2l+1)(1-|\eta_l|^2).$$

Therefore

$$\sigma_{abs} = \pi\lambda^2 \sum_{l=0}^{\infty}(2l+1)(1-|\eta_l|^2).$$

From these expressions it is seen that, for a given l-value, σ_{el}^l has maximum value $4\pi\lambda^2(2l+1)$ for $\eta = -1$ while the maximum absorption occurs when $\eta_l = 0$ giving $\sigma_{abs}^l = \pi\lambda^2(2l+1)$. The total cross-section

$$\sigma_{tot}^l = \sigma_{el}^l + \sigma_{abs}^l = \pi\lambda^2(2l+1)(|1-\eta_l|^2 + 1 - |\eta_l|^2)$$
$$= \pi\lambda^2(2l+1)(2 - \eta_l - \eta_l^*)$$
$$= \pi\lambda^2(2l+1)2(1 - \mathcal{R}(\eta_l))$$

which has a maximum value $4\pi\lambdabar^2(2l + 1)$ when $\eta_l = -1$, corresponding to the maximum of the elastic scattering and no absorption. A particularly simple application of these results gives the Black Nucleus approximation in which one assumes that η_l takes the values zero for $l \leqslant kR_N$ and unity for $l > kR_N$. This corresponds to the classical impact parameter description since $\hbar kR_N$ is the maximum angular momentum brought in by a particle which (classically) hits, and is absorbed by, the nucleus. Thus

$$\sigma_{abs}(BN) = \pi\lambdabar^2 \sum_{l=0}^{kR_N} (2l + 1) = \pi\lambdabar^2(kR_N + 1)^2 = \pi(R_N + \lambdabar)^2.$$

From the expression for elastic scattering $\sigma_{el}(BN)$ also has this value—a result which has been commented on previously (p. 7).

The scattering parameters at the nuclear surface

So far the scattering and absorption have been analysed in terms of parameters based on the wave function at infinity, and these must depend on boundary conditions at the surface of the nucleus. The two conditions, continuity of ψ and of its spatial derivatives at the nuclear surface, require that the (θ, φ) dependence be the same either side of the surface, which is equivalent to conservation of orbital angular momentum; their effect on the radial dependence can be conveniently combined to require that

$$\frac{\partial(\log \psi)}{\partial(\log r)} = \frac{r}{\psi}\frac{\partial \psi}{\partial r}$$

be continuous at the nuclear surface. If we restrict our attention to the $l = 0$ component of the wave (if $kR_N \ll 1$, then the other components at the nuclear surface are quite negligible anyway) the mathematics is made simpler by the fact that the form of the $l = 0$ wave function at infinity, namely

$$\psi = \frac{1}{2ikr}[\eta_0 e^{ikr} - e^{-ikr}],$$

persists right down to the nuclear surface. For neutral particles with other values of l this is not the case, nor is it so for charged particles; to cope with these cases it is necessary to know more about spherical Bessel functions and their complementary functions which are irregular at the origin, and the equivalent regular and irregular Coulomb wave-functions for charged particles. (The reader may be familiar with solving similar problems in electrostatics and magnetostatics by using the 'dipole at infinity' which gives a potential which is regular at the origin and the 'dipole at the origin' which gives a potential which is irregular at the origin.)

To revert to $l = 0$,

$$\log \psi = \log(\eta_0 e^{ikr} - e^{-ikr}) - \log r - \log 2ik$$

therefore

$$r \frac{d(\log \psi)}{dr} = \frac{r}{\psi} \frac{d\psi}{dr} = ikr \frac{\eta_0 e^{ikr} + e^{-ikr}}{\eta_0 e^{ikr} - e^{-kr}} - 1.$$

Putting $r = R_N$ and equating the resulting expression with its appropriate value just inside the nucleus leads to a solution for η_0. But to arrive at this appropriate value just inside the nucleus requires some drastic over-simplification of the problem. If we assume a square well potential then, just inside the well, the wave function will consist of the two components e^{-iKr}/r and e^{iKr}/r representing waves moving inwards and outwards respectively. Further, if we wish to consider a situation in which no absorption of particles occurs then the amplitudes of these waves must be equal. (Note that no absorption neglects the possibility of radiative capture occurring and this may be a very bad approximation for neutrons in the eV region.) It is tempting to assume that this form of the wavefunction persists all through the nucleus when the requirement that the wavefunction be finite at the origin determines the relative phase of the two components at the nuclear surface (they are obviously equal in amplitude). But this single-particle picture is an over-simplification of a nuclear state, as Bohr realised in his model; the single-particle state is mixed into the great number of states corresponding to multi-particle excitations. We therefore take the internal wavefunction in this channel and just inside the nuclear surface to be $\psi \propto [e^{i2\varphi} e^{iKr} - e^{-iKr}]/r$ where φ is an unknown function of the energy in the channel but not of the spatial coordinates. At the nuclear surface

$$\frac{r}{\psi} \frac{d\psi}{dr} = ikR_N \frac{e^{i2\varphi} e^{iKR_N} + e^{-iKR_N}}{e^{i2\varphi} e^{iKR_N} - e^{-iKR_n}} - 1$$

and the equation of continuity is therefore

$$e^{iKR_N} \frac{e^{i2\varphi} e^{iKR_N} + e^{-iKR_N}}{e^{i2\varphi} e^{iKR_N} - e^{-KR_N}} - 1 = ikR_N \frac{\eta_0 e^{ikR_N} + e^{-ikR_N}}{\eta_0 e^{ikR_N} - e^{-ikR_N}} - 1$$

Solving for η_0 gives

$$\eta_0 = \frac{e^{i2\varphi}(K + k)e^{iKR_N} + (K - k)e^{-iKR_N}}{e^{i2\varphi}(K - k)e^{iKR_N} + (K + k)e^{-iKR_N}} e^{-i2kR_N}$$

and

$$(1 - \eta_0) = e^{-i2kR_N}\left(e^{i2kR_N} - 1 - \frac{4ik \sin(KR_N + \varphi)}{2K \cos(KR_N + \varphi) - 2ik \sin(KR_N + \varphi)}\right)$$

$$= e^{-i2kR_N}\left(e^{i2kR_N} - 1 - \frac{2ik}{K \cot(KR_N + \varphi) - ik}\right)$$

from which $\sigma_{el} = \pi\lambdabar^2 \,|1 - \eta_0|^2$.

Hard sphere scattering and resonant scattering

The first two terms in the bracket above are of order $i2kR_N$ whereas the third term has the value 2 ($\gg 2kR_N$ in the present treatment) when $\cot(kR_N + \varphi) = 0$, falling to zero when $\cot(kR_N + \varphi) \to \infty$. But the wave function just inside the nucleus can also be expressed as $2i\sin(Kr + \varphi)e^{i\varphi}/r$ and therefore has the value zero at the nuclear surface when $\cot(KR_N + \varphi) \to \infty$. This situation therefore corresponds to a node at the nuclear surface for the external wave function also and the resulting scattering is therefore known as hard sphere scattering and has the cross-section

$$\sigma_{el} \text{ (hard sphere)} = \pi\lambdabar^2 \,|e^{i2kR_N} - 1|^2 = 4\pi\lambdabar^2 \sin^2(kR_N)$$

giving the result $4\pi R_N^2$ for $kR_N \ll 1$. The third term in the bracket gives resonant scattering. For $kR_N \ll 1$ we can neglect the hard sphere scattering term when near to the maximum (resonant) energy and arrive at

$$\sigma_{el}(\text{res}) = \pi\lambdabar^2 \left|\frac{-2ik}{K \cot(KR_N + \varphi) - ik}\right|^2$$

For $\cot(KR_N + \varphi) = 0$ at $\varphi = \varphi_{res}$ then $KR_N + \varphi_{res} = (2n + 1)\pi/2$. Assuming that a linear expansion in energy will suffice for φ near the resonance then

$$\varphi = \varphi_{res} + \frac{d\varphi}{d\varepsilon}(\varepsilon - \varepsilon_r) \quad \text{and} \quad \cot(kR_N + \varphi) \sim -\frac{d\varphi}{d\varepsilon}(\varepsilon - \varepsilon_r).$$

The square modulus term therefore becomes

$$\left|\frac{-2ik}{-K\frac{d\varphi}{d\varepsilon}(\varepsilon - \varepsilon_r) - ik}\right|^2 = \left|\frac{i2k/K\frac{d\varphi}{d\varepsilon}}{(\varepsilon - \varepsilon_r) + ik/K\frac{d\varphi}{d\varepsilon}}\right|^2$$

Putting $\Gamma_n = 2k/K(\mathrm{d}\varphi/\mathrm{d}\varepsilon)$ gives

$$\delta_{el}(\text{res}) = \pi\lambda^2 \frac{\Gamma_n^2}{(\varepsilon - \varepsilon_r)^2 - \frac{1}{4}\Gamma_n^2}$$

which is recognizable as the Breit–Wigner formula when only the entrance channel is available (because our 'neutrons' are spinless the statistical factor is unity). Thus the unknown phase function is a parameter which determines the neutron width of the resonance.

On average φ changes by π when ε changes by D (the mean level spacing) giving an average value for $\mathrm{d}\varphi/\mathrm{d}\varepsilon$ of π/D and an average Γ_n of

$$\frac{2k}{K}\frac{D}{\pi} = \frac{4k}{K}\frac{D}{2\pi} = T_0\frac{D}{2\pi}$$

where T_0 is the transmission coefficient at low energies and has appeared in other parts of this book. In practice one must suppose that φ does not increase linearly with energy but oscillates irregularly about this linear increase in such a way that $KR_N + \varphi = (2n+1)\pi/2$ at each resonant energy but the slope $\mathrm{d}\varphi/\mathrm{d}\varepsilon$ will fluctuate about the mean value π/D. However this simple treatment has produced the concept of resonant elastic scattering interfering with non-resonant (hard sphere) scattering and has given the resonant scattering the correct (Breit–Wigner) energy dependence. It has further shown that the width in a given channel is of the order of $1/2\pi$ times the local level spacing reduced by the transmission coefficient. The concept of absorption (i.e. nuclear reaction) can be introduced by amending the factor $e^{i2\varphi}$ to $Ae^{i2\varphi}$ where A is real and lies between 0 and 1. Other l-values can be considered, other particles (charged), and finally the particles can be given spin—but all these are achieved at the expense of greater mathematical complexity. Also the treatment here is limited to considering absorption of particles in the channel only to give the total reaction cross-section. To account for a given type of reaction requires a much greater degree of sophistication based on the concept of the Scattering Matrix which connects what is happening in one channel with another channel.

Bibliography

Other textbooks recommended for further reading, some of which have been referred to in the text, are:

*BLATT, J. M. and WEISSKOPF, V. (1952). *Theoretical nuclear physics.* Wiley, New York.

BOWLER, M. G. (1973). *Nuclear physics.* Pergamon Press, Oxford.

BURCHAM, W. E. (1963). *Nuclear physics, An introduction.* Longman, London.

*DIRAC, P. A. M. (1958). *Principles of quantum mechanics.* Clarendon Press, Oxford.

ENGE, H. (1966). *Introduction to nuclear physics.* Addison-Wesley, New York.

EVANS, R. D. (1955). *The atomic nucleus.* McGraw-Hill, New York.

FERMI, E. (1950). *Nuclear physics.* Chicago University press.

*HODGSON, P. E. (1963). *The optical model of elastic scattering.* Clarendon Press, Oxford.

KUHN, H. G. (1970). *Atomic spectra.* Longmans, London.

MARTIN, J. L. (1981). *Basic quantum mechanics.* Clarendon Press, Oxford.

PAUL, E. B. (1969). *Nuclear particle physics.* North-Holland, Amsterdam.

PAULING, L. and WILSON, E. B. (1935). *Introduction to quantum mechanics.* McGraw-Hill, New York.

PERKINS, D. H. (1972). *Introduction to high-energy physics.* Addison-Wesley, New York.

*PRESTON, M. A. (1962). *Physics of the nucleus.* Addison-Wesley, New York.

*SCHIFF, L. E. (1968). *Quantum mechanics.* McGraw-Hill, New York.

SEGRE, E. (1964). *Nuclei and particles.* Benjamin, New York.

Those indicated by an * are postgraduate in standard.

Answers to problems

1.1. $n \sim 3$.
1.4. $\varepsilon \sim 0.63$, $\gamma \sim 19.1$.
1.5. 3.7 fm.
2.3. +0.31 nm.
2.6. ~ 2.07 MeV.
3.4. $I \sim \frac{2}{3} I_{\text{rigid}}$.
3.6. $\lambda \sim 16$, 4.8, 6.3, 16 fm respectively.
4.4. ~ 3 s.
5.1. $\sim 6 \times 10^{-18}$ s.
6.2. 7.0191 a.m.u.
6.6. 8.5 μb.
7.3. 0.001 15.
7.5. (a) 2.4 MeV; (b) 14 keV in centre-of-mass system;
    (c) $\sim 1.6 \times 10^8$ K; (d) thermal spread $\sim \pm 100$ keV.

Physical constants and conversion factors

Avogrado constant	L or N_A	$6.022 \times 10^{23}\,\text{mol}^{-1}$
Bohr magneton	μ_B	$9.274 \times 10^{-24}\,\text{J T}^{-1}$
Bohr radius	a_0	$5.292 \times 10^{-11}\,\text{m}$
Boltzmann constant	k	$1.381 \times 10^{-23}\,\text{J K}^{-1}$
charge of an electron	e	$-1.602 \times 10^{-19}\,\text{C}$
Compton wavelength of electron	$\lambda_C = h/m_e c = 2.426 \times 10^{-12}\,\text{m}$	
Faraday constant	F	$9.649 \times 10^4\,\text{C mol}^{-1}$
fine structure constant	$\alpha = \mu_0 e^2 c/2h = 7.297 \times 10^{-3}\,(\alpha^{-1} = 137.0)$	
gas constant	R	$8.314\,\text{J K}^{-1}\,\text{mol}^{-1}$
gravitational constant	G	$6.673 \times 10^{-11}\,\text{N m}^2\,\text{kg}^{-2}$
nuclear magneton	μ_N	$5.051 \times 10^{-27}\,\text{J T}^{-1}$
permeability of a vacuum	μ_0	$4\pi \times 10^{-7}\,\text{H m}^{-1}$ exactly
permittivity of a vacuum	ε_0	$8.854 \times 10^{-12}\,\text{F m}^{-1}\,(1/4\pi\varepsilon_0 = 8.988 \times 10^9\,\text{m F}^{-1})$
Planck constant	h	$6.626 \times 10^{-34}\,J\,s$
(Planck constant)/2π	$\hbar$	$1.055 \times 10^{-34}\,\text{J s} = 6.582 \times 10^{-16}\,\text{eV s}$
rest mass of electron	m_e	$9.110 \times 10^{-31}\,\text{kg} = 0.511\,\text{MeV}/c^2$
rest mass of proton	m_p	$1.673 \times 10^{-27}\,\text{kg} = 938.3\,\text{MeV}/c^2$
Rydberg constant	$R_\infty = \mu_0^2 m_e e^4 c^3/8h^3 = 1.097 \times 10^7\,\text{m}^{-1}$	
speed of light in a vacuum	c	$2.998 \times 10^8\,\text{m s}^{-1}$
Stefan–Boltzmann constant	$\sigma = 2\pi^5 k^4/15h^3 c^2 = 5.670 \times 10^{-8}\,\text{W m}^{-2}\,\text{K}^{-4}$	
unified atomic mass unit (^{12}C)	u	$1.661 \times 10^{-27}\,\text{kg} = 931.5\,\text{MeV}/c^2$
wavelength of a 1 eV photon		$1.243 \times 10^{-6}\,\text{m}$

$1\,\text{Å} = 10^{-10}\,\text{m}$; $1\,\text{dyne} = 10^{-5}\,\text{N}$; $1\,\text{gauss(G)} = 10^{-4}\,\text{tesla(T)}$;
$0°\text{C} = 273.15\,\text{K}$; $1\,\text{curie(Ci)} = 3.7 \times 10^{10}\,\text{s}^{-1}$;
$1\,\text{J} = 10^7\,\text{erg} = 6.241 \times 10^{18}\,\text{eV}$; $1\,\text{eV} = 1.602 \times 10^{-19}\,\text{J}$; $1\,\text{cal}_{\text{th}} = 4.184\,\text{J}$;
$\ln 10 = 2.303$; $\ln x = 2.303 \log x$; $e = 2.718$; $\log e = 0.4343$; $\pi = 3.142$

Index

thermal neutron capture 144
transmission coefficient 75

uncertainty principle 3, 13, 43, 49, 76

X-rays
   atomic 4
   muonic 4
   K-mesic 9

yrast states 133

Oxford Physics Series

£9-95

Oxford Physics Series

General Editors

WITHDRAWN

E. J. BURGE   D. J. E. INGRAM   J. A. D. MATTHEW